T0073060

Multiple Access Technology Towards Ubiquitous Networks

Neng Ye · Xiangming Li · Kai Yang · Jianping An

Multiple Access Technology Towards Ubiquitous Networks

Overview and Efficient Designs

 Springer

Neng Ye
Beijing Institute of Technology
Beijing, China

Kai Yang
Beijing Institute of Technology
Beijing, China

Xiangming Li
Beijing Institute of Technology
Beijing, China

Jianping An
Beijing Institute of Technology
Beijing, China

ISBN 978-981-19-4024-8   ISBN 978-981-19-4025-5 (eBook)
https://doi.org/10.1007/978-981-19-4025-5

This Springer imprint is published by the registered company Springer Nature Singapore Pte Ltd.
The registered company address is: 152 Beach Road, #21-01/04 Gateway East, Singapore 189721, Singapore

Preface

The future wireless communication networks are expected to provide ubiquitous massive coverage to meet the requirements of diversified applications. With an exponential increase of the number of users and devices, it is challenging to establish fast and reliable connections in the ubiquitous network. As the core technology in the evolution of wireless communication systems, multiple access technology can enable effective massive connections and has become the prominent research trend for ubiquitous networks. The goal of this book is to provide readers with a comprehensive overview of the state-of-the-art multiple access technologies for ubiquitous network, with a focus on the novel ubiquitous multiple access technologies, the signal construction techniques of multiple access and the enhanced multiple access transceivers using Artificial Intelligence (AI). For each of these topics, this book has tried to provide an advanced introduction, blending the basic multi-user information principles with the advanced multiple access models and novel application scenarios. In addition, this book has provided elaborate simulation results for each topic to verify the feasibility of the corresponding schemes.

In particular, to have a comprehensive view for the application scenarios of the multiple access technology, this book discusses the evolution and deployment of multiple access in 5G and beyond, followed by the emerging multiple access technologies developed for the ubiquitous non-terrestrial networks. Facing the challenges of massive connections in ubiquitous networks this book investigates the effective signal construction techniques of multiple access, including constellation design and rate splitting. Moreover, the enhancement of multiple access transceivers using AI is presented. Specifically, this book resorts to AI for constructing unified optimization framework and approaching the performance limit of multiple access system, and enhances grant-free multiple access to match the features of Internet of Things (IoT) with deep learning. This book systematically describes the theoretical framework and physical layer technologies of ubiquitous access, which can reflect the application prospect for future ubiquitous networks.

We believe that this book can provide useful insights for the theory and method of ubiquitous multiple access, and display the wide application of ubiquitous networks in the future 6G. This book can be used as a reference for graduate students, researchers, and engineers in the field of wireless communications. We do hope that the valuable time devoted to this book will bear fruit in stimulating interest in the study of multiple access technologies.

Beijing, China   Neng Ye
                                                                                                   Xiangming Li
                                                                                                     Kai Yang
                                                                                                  Jianping An

Acknowledgements The works in this book have been supported by the National Natural Science Foundation of China under Grants 62101048 and 62171030.

Contents

Acronyms

3GPP	Third Generation Partnership Project
A2G	Air-to-ground
ACRDA	Asynchronous contention resolution diversity slotted ALOHA
ACS	Asymmetric chirp signal
AI	Artificial intelligence
ARM	Adaptive random-selected multi-beamforming
AWGN	Additive white Gaussian noise
BA	Buffering-aided
BC	Broadcast channel
BER	Bit error rate
BS	Base station
BTCs	Block turbo codes
CCC	Constellation-constrained capacity
CDG	Compressive data gathering
CE	Cross-entropy
CoAP	Constrained limited application protocol
CP	Cyclic-prefix
CRDSA	Resolution diversity slotted ALOHA
CS	Compressive sensing
CSA	Coded slotted ALOHA
CSI	Channel state information
CSS	Chirp-spread spectrum
CTU	Contention-based unit
D2D	Device-to-device
DAE	Deep auto-encoding
DAS	Delay-aware selection
DAUS	Delay sensing UAV selection
DC	Difference of convex
DE	Density evolution
DL	Deep Learning
DL-PA	Deep learning-based long-term power allocation

DNN	Deep neural network
DSA	Diversity slotted ALOHA
E2E	End-to-end
EAS	Energy-aware selection
EAUS	Energy sensing UAV selection
EE	Energy efficiency
EM	Expectation-maximization
eMBB	Enhanced mobile broadband
EPA	Estimation propagation algorithm
FEC	Forward error-correction
FO	Frequency offset
FTPA	Fractional transmit power allocation
FTUS	Fair weighing UAV selection
G2G	Ground-to-ground
GEO	Geostationary earth orbit
GOCA	Group orthogonal coded access
HAP	Hybrid access point
HARQ	Hybrid automatic repeat request
IC	Interference channel
ICI	Inter-cell interference
ICNN	IC-enabled DNN
IDMA	Interleave-division multiple-access
IGMA	Interleave-grid multiple access
IoT	Internet of things
IRSA	Irregular repetition slotted ALOHA
KKT	Karush-Kuhn-Tucker
KL	Kullback-Leibler
LAN	Local area network
LCRS	Low coding rate spreading
LEO	Low Earth orbit
LLRA	Low-latency routing algorithm
LPWAN	Low power wide area network
LSSA	Low code rate and signature based shared access
MABs	Multiple access blocks
MAC	Multiple access channel
MAP	Maximum a posteriori
MASs	Multiple access signatures
MBB	Mobile broadband
MCSs	Modulation and coding schemes
MER	Message error rate
MF	Match-filter
MI	Mutual information
ML	Maximum-likelihood
mMTC	Massive machine-type communication
MPA	Message passing algorithm

MUD	Multi-user detection
MUI	Multi-user interference
MUSA	Multi-user sharing access
MUST	Multi-user sharing technology
NCC	Non-orthogonal cover codes
NCMA	Non-orthogonal coded multiple access
NFV	Network function virtualization
NOCA	Non-orthogonal coded access
NOMA	Non-orthogonal multiple access
NOSA	Non-orthogonal slotted ALOHA
OCCs	Orthogonal cover codes
OFDM	Orthogonal frequency division multiplexing
OFDMA	Orthogonal frequency-division multiple access
OMA	Orthogonal multiple access
PAPR	Peak to average power ratio
PDF	Probability density function
PDMA	Pattern division multiple access
PF	Proportional fairness
PIC	Parallel interference cancellation
PMF	Probability mass function
RA	Random access
RAC	Random access channel
RACH	Random access channel
RAMA	Rate-adaptive multiple access
RAR	Random access response
RB	Resource block
RDMA	Repetition division multiple access
REs	Resource elements
RHS	Right-hand side
RLNC	Random linear network coding
ROC	Receiver operating characteristic
RS	Rate splitting
RSMA	Resource spread multiple access
SA	Slotted ALOHA
SC	Superposition coding
SC-FDMA	Single-carrier frequency division multiple access
SCMA	Sparse code multiple access
SCS	Symmetric chip signal
SCSS	Symmetry chirp spread spectrum
SDN	Software defined network
SE	Spectral efficiency
SGD	Stochastic gradient decent
SIC	Successive interference cancellation
SIN	Space information network
SINR	Signal to interference and noise ratio

SJD	Successive joint decoding
SM	Spatial modulation
SR	Scheduling request
SRRS	Super-imposed radio resource sharing
SSA	Spread slotted ALOHA
SSMA	Short sequence spreading-based multiple access
TBSs	Transmission block sizes
TDL	Tapped-delay-line
TDMA	Time division multiple access
TMs	Transmission modes
TO	Timing offset
TP	True positive
TPA	Transmission power allocation
UAV	Unmanned aerial vehicle
UCI	Uplink control information
UEP	Unequal protection property
UNB	Ultra narrow band
VA	Variational approximation
VAE	Variational auto-encoder
VA-M	Variational approximation-maximization
VMF	Von-mises-fisher
VPs	Variational parameters
WBE	Welch-bound-equality

Chapter 1
Introduction

Abstract This chapter first introduces the necessity of enhancing multiple access technology for ubiquitous networks. Then, the evolution of multiple access is briefly introduced and the aspects of enhancement methods of multiple access including signal construction and AI-based transceiver design are discussed. Finally, the organization of this book is presented.

1.1 Background

How to establish stable and fast wireless connections of multiple mobile users is one of the key issues in a wireless communication system. Multiple access technology, which constructs multiple interference-limited single-user transmission channels by dividing wireless resources in time, frequency, space, code and power domain, is the core solution to address the key issues mentioned above. As a matter of fact, multiple access technology has been the key enabler for the intergenerational evolution of wireless communication system from 1G (1980s) to 5G (2020s) [1].

Wireless communication system towards 2030 is expected to break the limitations of time/space and realize ubiquitous interconnections in multiple (air/space/ground) domains [2]. While conventional cellular-based networks mainly focus on the terrestrial hotspot coverage, the spaceborne and airborne platforms in the ubiquitous networks can directly serve widely distributed user equipments [3–5]. These platforms can supply uninterrupted and undifferentiated communication for users, due to their full coverage, all-time work, robustness to damage, flexibility and reliability. Therefore, ubiquitous networks will provide seamless wireless connections and tend to have a profound impact in the future 6G on coverage and connection density. At present, radio access technology for ubiquitous networks has been a focus of international standardization organizations such as 3GPP and ITU [6, 7].

Different from terrestrial networks, ubiquitous networks have the characteristics of massive access users, large propagation distances, high channel dynamics, dynamic topology, and limited power budget. Due to the above-mentioned differences, conventional multiple access technologies for terrestrial networks are not suitable for ubiquitous access scenarios and requirements. To this end, it is necessary to study advanced multiple access technology to improve the connectivity, spectral efficiency, and reliability.

Inspired by the above stringent requirements, this book focuses on the state-of-the-art multiple access technologies for ubiquitous networks. Specifically, this book firstly investigates novel ubiquitous multiple access technologies for beyond 5G and space-air-ground integrated networks in 6G. Then, the signal construction techniques of multiple access are studied, including constellation design and rate splitting. Moreover, the enhancement of multiple access transceivers using artificial intelligence (AI) is studied. In summary, this book systematically describes the theoretical framework and physical layer technology of ubiquitous access, which can deepen the understanding of the theory and method of ubiquitous multiple access, and promote the wide application of ubiquitous networks in the future 6G.

1.2 Evolution of Multiple Access Technology

Multiple access technology originates from multi-user information theory [8]. As early as the 1970s, the research of multi-user information theory, i.e. network information theory, theoretically pointed out that any point in the capacity domain of multiple access channels and degraded broadcast channels can be achieved by superposition coding (SC), rate splitting (RS) and successive interference cancellation (SIC) reception [9, 10].

The orthogonal multiple access technology applied to the 1G-5G system considers the orthogonal division of wireless resources and cannot reach the outer boundary of multi-user capacity region, and suffers from limited access capability. Therefore, it is necessary to study efficient multiple access technologies for ubiquitous networks. Recently, the idea of non-orthogonal multiple access (NOMA) has been proposed to achieve the entire capacity region of multiple access channel [11]. NOMA allows the superposition transmissions of multiuser signals with controllable mutual interference. By deploying advanced multiuser detector, NOMA significantly enhances the connectivity, improves the spectral efficiency and simplifies the signaling interactions compared with its orthogonal counterpart.

The earliest application of the non-orthogonal paradigm in multi-user information theory to actual systems can be traced back to Gianluca Mazzini's paper "Power Division Multiple Access" in 1998 [12]. The key is to explicitly introduce the power dimension and deploy cascaded detectors at the receiving end. In 2008, the research group led by NTT docomo put forward a method to overlap and reuse the uplink resources in the patent "Physical Resource Allocation Method, Device, Data Receiving Method and Receiver". It is proposed that the natural near-far effect of CDMA can be used to pair users with significantly different equivalent receiving powers and realize uplink transmission with the same time-frequency physical resources, and then distinguish users thruogh SIC receivers. This patent is the basic composition patent of non-orthogonal multiple access in the industry.

1.3 Signal Construction for Multiple Access Technology

Multiple access technology allows multiple users to perform superimposed transmission on wireless resources through specific transceiver design. This ensures a good compromise between multiplexing gain and transmission reliability under controllable inter-user interference. In the general model of multiple access systems, each user uses a specific transmitter and maps the source message into a coded signal based on its specific access feature fingerprint. The coded signals of the multiple users are then superposed in the wireless channel, and finally the receiver recovers the source message using multi-user detection algorithm. Typically, multiple access transmitters map the source messages (or bits) to complex modulation symbols (or symbol sequences); and multi-user receivers then map the received signal to an estimated source message. Therefore, design of transmission signal becomes a major challenge in the research of multiple access technology [13].

In the design of transmitters, due to the difficulty of directly indicate the overall system performance of multiple access, it is necessary to introduce some easy-to-characterized indicators, such as Euclidean distance or channel capacity measurement. For practical implementation, finite-alphabet signal should be transmitted. To this end, signal design based on constellation-constrained capacity (CCC) needs to be studied [14]. After fixing the single-user constellation diagram, it is a simple and effective way to improve the system capacity by adjusting the rotation angle of the constellation diagram. However, the existing literature mainly maximizes the CCC by assuming the ideal maximum likelihood receiver, which leads to high receiving complexity. Therefore, it is necessary to enhance the performance of multiple access system with practical receiver via constellation rotation.

Another promising technique in multiple access signal design is rate splitting (RS) [15]. RS is considered as a promising physical layer transmission mode for non-orthogonal transmission, interference management and multiple access strategies in 6G. By splitting the user message into multiple independently coded signal layers at the transmitter and partially decoding the interference and treating the remainder of the interference as noise, the RS is able to gently bridge two extreme interference management strategies, namely, interference is treated as noise (commonly used in 4G/5G multi-user/mass/mmWave MIMO) and fully decode interference (in non-orthogonal multiple access). RS can also effectively improve the robustness of grant-free access systems by assigning different reliability to different data streams. Therefore, it is promising to design a practical multiple access scheme to address the unpredictable interference in grant-free access and to fully utilize the underlaid physical channel.

1.4   AI-Enhanced Multiple Access Technology

With the beginning of the 2020s, the existing multiple access technology system can no longer meet the expectations of the future intelligent society for ubiquitous wireless access. For the wireless access of user equipment, one of the core requirements of the future smart society is to use a unified technical framework to realize intelligent ubiquitous connections—covering multiple meanings such as the number of massive devices, extensive space-time characteristics, different business attributes, and diversified performance indicators. This puts forward higher expectations for the next-generation multiple access technology.

With the help of information theory and signal processing methods, we have been able to approach the performance limit of the point-to-point communication system. However, these methods face difficulties in modeling and optimizing the complicated multi-user systems towards end-to-end transmission performance. Therefore, a new research paradigm is required to further optimize multiple access technology.

Deep Learning (DL), which automatically extracts the distributed features of the signals using deep models, provides unified signal processing architecture, universal function approximation ability and data-driven end-to-end optimization capability [16]. The recent breakthrough of DL and its positive applications to wireless communications have paved the way to tackle the above challenges of multiple access technology. At present, AI technology represented by deep learning has been closely coupled with mobile communication [17–20]. With the update iteration of hardware semiconductor technology including cloud graphics processor and terminal neural processing unit [21], it is expected to jointly enable a new intelligent wireless air interface. However, the current DL approach for physical-layer enhancement is still in its infancy. For example, many researches directly reuse the existing deep learning models of AI, which lack the targeted design of communication systems. And most of them only consider point-to-point communication scenarios, which lack a unified end-to-end deep learning framework for multi-access systems. Moreover, the potential superposition in multiple access technology further leads to exponentially increased combinations of the feature, which requires the sophisticated design of DL.

1.5   Organization

To have a comprehensive view of the multiple access technology, this book discusses the the multiple access towards different scenarios in Chaps. 2 and 3, presents effective signal construction methods of multiple access in Chaps. 4 and 5, and multiple access technologies enhanced by AI and DL in Chaps. 6 and 7. The rest of this book is organized as follows.

Chapter 2 discusses the practical deployment of multiple access in 5G and beyond. As a vital variant of multiple access, non-orthogonal multiple access (NOMA) plays an important role in the standardization process of 5G, with regards to spectral efficiency, reliability, low latency and peak data rate. This book presents a comprehensive review on recent progress of multiple access technology in 5G and beyond, especially NOMA technologies.

Chapter 3 discusses multiple access technology used in non-terrestrial wireless communication systems for ubiquitous networks. Non-terrestrial networks here help to break the location constraints that existing terrestrial IoT meets. This book provides a general look on the non-terrestrial network, including application scenarios, technical proposals, key techniques and potential research directions.

Chapter 4 investigates the constellation rotation technique for enhancing the performance of uplink multiple access network. A successive interference cancellation (SIC) receiver is applied to achieve better capability, which is also the basic receiver used in this proposed method. In this chapter, the best value of rotation angle is obtained by maximizing the entropy of Gaussian mixture model and then used to characterize the receiving signal. Lower bit error rate and larger capacity compared to conventional multiple access can be found in the proposed structure.

Chapter 5 investigates rate-adaptive multiple access for uplink grant-free transmission. Grant-free transmission helps simplify the signaling procedure through uplink instant transmission, while in the other hand its collision problem causes smaller data throughput and worse outage performance. In this chapter, a rate-adaptive multiple access (RAMA) scheme is put forward to tackle the problem. The corresponding receiver applied with successive interference cancellation algorithm is introduced to detect multiple data streams. Compared to conventional grant-free scheme, RAMA scheme can achieve higher average throughput and lower outage performance.

Chapter 6 investigates AI-aided multiple access for end-to-end optimization. Deep learning is used to further approach the performance limit of NOMA. This chapter regards the overlapped transmissions in NOMA as multiple distinctive but correlated learning tasks, and then puts forward a unified multi-task deep neural network (DNN) framework for NOMA, namely DeepNOMA. Compared to the conventional methods, DeepNOMA can achieve higher transmission accuracy and lower computational complexity simultaneously under various channel models.

Chapter 7 investigates grant-free multiple access based on deep learning in a specific scenario, i.e., tactile Internet of Things (IoT). The benefit of grant-free access and non-orthogonal transmissions are jointly exploited to achieve low latency massive access. However, reliability seems to be reduced due to the random interference. This chapter formulates a variational optimization problem to improve the reliability of grant-free NOMA. DNN is used to tackle the proposed problem and the training process matches the highly automatic applications in tactile IoT. Significant reliability gain can be found in the proposed scheme.

Chapter 8 summarizes this book and discusses the future directions of enhancing multiple access technology for ubiquitous networks.

References

1. Z. Ding, Y. Liu, J. Choi et al., Application of non-orthogonal multiple access in LTE and 5G networks. IEEE Commun. Mag. **55**(2), 185–191 (2017)
2. Z. Zhang, Y. Xiao, Z. Ma et al., 6G wireless networks: vision, requirements, architecture, and key technologies. IEEE Veh. Technol. Mag. **14**(3), 28–41 (2019)
3. G. Giambene, S. Kota, P. Pillai, Satellite-5G integration: a network perspective. IEEE Netw. **5**, 25–31 (2018)
4. S. Cioni, R. De Gaudenzi, O. Del Rio Herrero, N. Girault, On the satellite role in the era of 5G massive machine type communications. IEEE Netw. **32**, 54–61 (2018)
5. W. Chien, C. Lai, M.S. Hossain, G. Muhammad, Heterogeneous space and terrestrial integrated networks for IoT: architecture and challenges. IEEE Netw. **33**, 15–21 (2019)
6. 3gpp RAN 1. http://www.3gpp.org/
7. Y. Chen et al., Toward the standardization of non-orthogonal multiple access for next generation wireless networks. IEEE Commun. Mag. **33**, 19–27 (2018)
8. A.A. El Gamal, Y.H. Kim, *Network Information Theory* (Cambridge University Press, 2011), pp. 151–158
9. T. Cover, R. McEliece, E. Posner, Asynchronous multiple-access channel capacity. IEEE Trans. Inf. Theory **27**, 409–413 (1981)
10. S. Verdu, The capacity region of the symbol-asynchronous Gaussian multiple-access channel. IEEE Trans. Inf. Theory **35**, 733–751 (1989)
11. K. Yang, N. Yang, N. Ye et al., Non-orthogonal multiple access: achieving sustainable future radio access. IEEE Commun. Mag. **57**(2), 116–121 (2019)
12. G. Mazzini, Power division multiple access. Univ. Pers. Commun. **1**, 543–546 (1998)
13. Q. Zhang, H. Guo, Y.C. Liang et al., Constellation learning based signal detection for ambient backscatter communication systems. IEEE J. Sel. Areas Commun. **37**(2), 452–463 (2018)
14. J. Harshan, B.S. Rajan, On two-user Gaussian multiple access channels with finite input constellations. IEEE Trans. Inf. Theory **57**(3), 1299–1327 (2011)
15. B. Rimoldi, R. Urbanke, A rate-splitting approach to the Gaussian multiple-access channel. IEEE Trans. Inf. Theory **42**(2), 364–375 (1996)
16. Y. LeCun, Y. Bengio, Deep learning. Nature **521**(7553), 436–444 (2015)
17. X. You, C. Zhang, X. Tan et al., AI for 5G: research directions and paradigms. Sci. China Inf. Sci. **62**(2), 1–13 (2019)
18. Z. Qin, H. Ye, G.Y. Li et al., Deep learning in physical layer communications. IEEE Wirel. Commun. **26**(2), 93–99 (2019)
19. T. O'Shea, J. Hoydis, An introduction to deep learning for the physical layer. IEEE Trans. Cogn. Commun. Netw. **3**(4), 563–575 (2017)
20. T.J. O'Shea, T. Roy, T.C. Clancy, Overtheair: deep learning based radio signal classification. IEEE J. Sel. Top. Signal Process. **12**(1), 168–179 (2018)
21. J.D. Owens, M. Houston, D. Luebke et al., GPU computing. Proc. IEEE **96**(5), 879–899 (2018)

Chapter 2
Multiple Access Towards 5G and Beyond

Abstract In this chapter, we discuss the practical deployment of multiple access in 5G and beyond. Section 2.1 introduces the motivation of practically applying multiple access in wireless communication systems and present researches of multiple access technology. Section 2.2 investigates the typical technologies of multiple access in the existing researches. Section 2.3 presents the motivation and procedures of grant-free multiple access technology, including the transmission and reception techniques. Section 2.4 discusses the key issues raised by multiple access technology in the practical implementation process. Section 2.5 gives a conclusion of this chapter.

2.1 Introduction

In the past several decades, the wireless communication system has evolved from the first generation, an analog communication network which only transfers voice messages, to LTE networks, which satisfies the great demands on mobile broadband data transmissions. Recently, the development of 5G has raised new challenges with respect to peak data rate, user experience data rate, spectral efficiency (SE), energy efficiency (EE), massive connectivity, low latency, and ultra reliability, etc.

Non-orthogonal multiple access (NOMA) technologies have been recognized by both industry and academia as one promising tendency and progress, ever since the deployment of orthogonal frequency-division multiple access (OFDMA) in LTE, to meet the wide-ranging requirements for 5G and beyond under the strict constraint of the limited radio resources [1]. In the past several decades, the wireless communication system has evolved from the first generation, an analog communication network which only transfers voice messages, to LTE networks, which satisfies the great demands on mobile broadband data transmissions. Recently, the development of 5G has raised new challenges with respect to peak data rate, user experience data rate, spectral efficiency (SE), energy efficiency (EE), massive connectivity, low latency, and ultra reliability, etc.

The idea of NOMA can trace back to the information-theoretic researches about multi-user information theory [2]. In downlink broadcast channel (BC), superposition coding and successive interference cancellation (SIC) receiving are employed to

N. Ye et al., *Multiple Access Technology Towards Ubiquitous Networks*,
https://doi.org/10.1007/978-981-19-4025-5_2

approach the entire capacity region of BC. Meanwhile, in uplink multiple access channel (MAC), the signals of different transmitters are overlapped and SIC receiver is applied to achieve the corner points of MAC capacity region. In 1990s, multiple access protocols, which exploited the differences between the power levels of the received packets, were proposed and studied by Shimamoto [3], Pedersen [4], and Mazzini [5], respectively. In 2008, Yan and Li in [6] proposed a superimposed radio resource sharing (SRRS) scheme which utilizes the near-far effect to enhance the uplink throughput performance. SRRS superimposes different uplink data streams on the same radio resources and applies SIC at the receiver, which can be regarded a prototype of NOMA,

Despite all the related researches, NOMA is still not commercialized in the past decades due to the concern of high computational complexity of SIC-type receiver. However, the rapid growth of processing power of the microprocessors in these years has provided an opportunity to the standardization and commercialization of NOMA technologies. Recently, downlink non-orthogonal transmissions, featured by multi-user sharing technology (MUST), are specified in LTE Release-14 in 2017. Towards the evolution of 5G, the industrial community has proposed dozens of NOMA schemes as candidate multiple access technologies. In the meantime, a study item of NOMA, actuated by its potential advantages, has been approved by 3GPP RAN plenary [7] in March 2017, which promotes the standardization of NOMA in 5G.

The core idea of NOMA is to multiplex different data streams over the same radio resources and employ multi-user detection algorithm at the receiver to recover multiple users' signal streams. The major design target of NOMA is to introduce controllable mutual interference among users to achieve a fine tradeoff between multiplexing gain and detection reliability. According to both theoretical and numerical analysis, NOMA superpasses OMA with respect to SE, EE, and connectivity [8–11]. To grasp the development of NOMA technologies, some published review papers have presented different aspects of NOMA. We summarize the main contributions of these articles in Table 2.1.

Different from the existing literatures, this chapter presents a comprehensive review about the recent progress of NOMA proposed in the standardization process of 3GPP towards 5G, including candidate NOMA schemes and multi-user receiving technologies. Meanwhile, we also survey the state-of-art grant-free NOMA schemes, which are expected to satisfy the massive connectivity and high EE requirements in massive machine-type communication (mMTC) scenario. Additionally, we discuss the implementation issues about NOMA. The contributions of this survey is summarized in the following four aspects:

- A comprehensive survey about the candidate NOMA schemes proposed in 3GPP, as well as the promising multi-user detection methods. NOMA schemes are categorized into bit-level and symbol-level schemes for illustrations, according to the agreements in 3GPP [18].

Table 2.1 Summery of existing surveys about NOMA

Survey	Scope	Contributions
Islam et al. [12]	Power-domain NOMA	A comprehensive survey about power domain-NOMA, as well as the related designs
Dai et al. [13]	NOMA schemes	Review of power-domain and code-domain NOMA schemes
Tao et al. [14]	NOMA schemes and waveforms	Review of some NOMA schemes and non-orthogonal waveforms
Wang et al. [15]	NOMA schemes	Review of some NOMA schemes towards 5G, as well as the application scenarios and typical receivers
Ding et al. [16]	Theoretical analysis of NOMA	Review of the theoretical analysis about power-domain NOMA and cognitive radio inspired NOMA
Wei et al. [17]	Downlink NOMA	Industrial view about downlink NOMA in 5G

- The motivation and main idea of grant-free NOMA are presented in this survey. In addition to the grant-free procedures, this survey also introduces the typical grant-free NOMA schemes, as well as the detection algorithms.
- The implementation issues about NOMA, especially grant-free NOMA, are discussed, with respect to resource allocation, procedures, and physical layer signals.
- The future research challenges related to NOMA are identified, including physical layer enhancement, cross-layer design, applications of NOMA in new scenarios, and the joint design of NOMA with other technologies.

2.2 Typical Multiple Access Technologies

The industrial community has proposed plenty of NOMA schemes to meet the diversified requirements towards 5G. Until 3GPP TSG-RAN WG1 (RAN-1) #86, at least 15 candidate NOMA schemes have been proposed for 5G new radio (NR) [19–32]. On RAN-1 #86b, a general framework of NOMA schemes is agreed [18], which helps to categorize existing operations in NOMA schemes into bit-level operations and symbol-level operations, as shown in Fig. 2.1. Note that several component operations may be simultaneously adopted in the future 3GPP Release-15 to satisfy the requirements of different application scenarios.

Correspondingly, the proposed NOMA schemes can also be categorized into two classes, namely bit-level NOMA and symbol-level NOMA, where the bit-level NOMA focuses on the design related to channel coding and bit-level interleaving, while the symbol-level NOMA mainly lays emphasis on symbol spreading and mapping. According to the above classifications, we summarize the state-of-the-

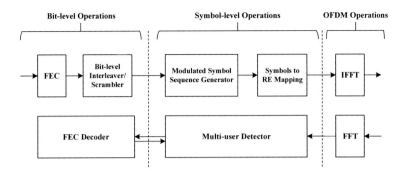

Fig. 2.1 A unified structure of NOMA technologies

art NOMA schemes in Table 2.2 and give a comprehensive analysis in the following. Meanwhile, the detection algorithms designed for these NOMA schemes are analyzed afterwards.

2.2.1 Bit-Level Non-orthogonal Multiple Access

Bit-level NOMA schemes exploit the low-rate forward error-correction (FEC) codes to enhance the detection accuracy, and/or take the advantage of user-specific interleaving to whiten the multi-user interference (MUI). In the following, we analyze several typical bit-level NOMA schemes include power-domain NOMA (PD-NOMA), low coding rate spreading (LCRS), low code rate and signature based shared access (LSSA), interleave-division multiple access (IDMA), and interleave-grid multiple access (IGMA).

2.2.1.1 PD-NOMA

PD-NOMA [27] multiplexes the users in power domain, and applies the iteration-based SIC receiver to detect multiple signal streams at the receiver [33, 34]. In each iteration of SIC receiving, the MUI is regarded as thermal noise, which suggests that the user de-multiplexing could be implemented by generating a large power difference among the multiplexed users. According to the simulation results, PD-NOMA can improve the resource utilization efficiency in both uplink and downlink [34]. Meanwhile, PD-NOMA can maintain low peak to average power ratio (PAPR) if single carrier property is kept [35]. In addition, PD-NOMA does not depend on the information of instantaneous channel state information (CSI) of frequency-selective fading. Therefore, no matter the user mobility or CSI feedback latency, a robust performance gain in practical wide area deployments can be expected.

Table 2.2 A summary of NOMA schemes

NOMA scheme	Key technical point			Main advantage
IDMA	Low coding rate	Low rate FEC code	Bit-level Interleaving	Randomized the mutual interference
IGMA		Low rate FEC code or moderate one with repetition	Bit-level Interleaving (permutation matrix)	Sparse grid Mapping
LSSA		Low rate FEC code or moderate one with repetition	User-specific bit-level interleaving/permutation pattern	Large number of signatures
LCRS		Low rate FEC code and repetition	Bit-level spreading	Large coding gain
SCMA	Short low density spreading	Multi-dimensional modulation		Signal space diversity gain
PDMA		Irregular LDS		Irregular protection
LDS-SVE		LDS & User signature vector extension (SVE)		Higher diversity
MUSA	Short dense spreading (low cross-correlation sequence)	Short complex spreading sequence		Easy to generate & Large number
NCMA		NCC obtained by Grassmannian line packing problem		Optimal non-orthogonal sequence
NOCA		Zadoff-Chu sequence		Easy to generate, low PAPR
SSMA		Orthogonal or quasi-orthogonal codes		
GOCA	Long spreading/scrambling sequence	Group-based orthogonal/non-orthogonal sequences		Inter-group orthogonality
RDMA		Cyclic shift based time-frequency repetition		Easy implementation
RSMA		Low cross-correlation Sequence scrambling		Fit for asynchronized scenario
RSMA(single tone)	Single carrier (similar to CDMA), low PAPR modulation			Extended coverage, and low PAPR for uplink

The major design aspect related to PD-NOMA is the resource allocation, including user association, radio resources assignment, and power allocation [36]. However, solving the resource allocation problem in one shot would be non-trivial. Therefore, this problem is usually decoupled into two subproblems, i.e., user scheduling and power allocation, respectively. In PD-NOMA, the users with large channel gain difference (e.g., large path-loss difference) are normally paired to enhance SE performance [37]. However, this simple criterion may cause unfairness in system-level deployment. Proportional fairness (PF) based scheduling [38], which simultaneously optimizes the user fairness and system throughput, is a practical user scheduling technology for PD-NOMA. The PF metric, calculated by dividing the instantaneous signal to interference and noise ratio (SINR) with the average data rate over the past period, is maximized during the user scheduling stage [39]. In uplink, user scheduling should consider the single-carrier frequency division multiple access (SC-FDMA) where the subcarriers are distributed continuously to overcome the PAPR problem. One low complexity heuristic method based on greedy subband widening is proposed in [40] for practical deployment.

In the meantime, there have been abundant literatures which address the power allocation problem of PD-NOMA [12]. Due to the non-convexity of the power allocation problem, advanced optimization techniques are usually employed to optimize the system throughput, reliability, and/or connectivity. In [41], the maximization of PF metric is presented. At first, the optimal power allocation of MAC is calculated iteratively. Then the optimization results can be converted to BC based on uplink-downlink duality. Several water-filling based methods are summarized and further studied in [42], where a weighted water-filling method is proposed in presence of user priority. In [43], an iterative sub-optimal power allocation algorithm based on difference of convex (DC) programming is presented. The readers may refer to [12] for an extensive review about resource allocation algorithms in PD-NOMA.

Nevertheless, the existing methods may still be complex for system-level deployment. Hence, several power allocation methods are proposed by industrial community to enable efficient and practical applications. When the users have been paired into groups, one option is to apply the pre-defined power allocation ratios to different users as done in [44]. An alternative method is to choose one option that can maximize the PF metric out of several options, e.g. for two-user NOMA, the options of the power ratio can be [0.2, 0.8] and [0.3, 0.7], which is also termed fixed transmission power allocation (TPA). Since the indexes of TPA can be pre-defined, TPA can effectively decrease the amount of downlink signaling related to PD-NOMA. Another commonly used method is the fractional transmit power allocation (FTPA) inspired from the transmission power control used in the LTE uplink [39]. In FTPA, the users with poorer channel conditions are allocated with more power to partially compensate the channel loss. In the above FTPA, the related parameters can be optimized via system-level simulation. After the resource allocation stage, a sophisticated design of the constellations, e.g., constellation rotation [45], may provide additional gain in enhancing the detection accuracy.

When multi-cell or dense-network scenario is considered [46], the uplink PD-NOMA would increase the inter-cell interference (ICI) because multiple users are allowed to transmit on the shared carriers. Therefore, user association and ICI-aware power allocation should also be studied to control the transmission power and avoid causing severe ICI to the neighboring cells [47].

2.2.1.2 IDMA

In addition to the power domain multiplexing, the users can also be distinguished if they have different interleaving patterns, which is exploited in interleave-division multiple-access (IDMA). IDMA is initially proposed by Li et al. [48] to enhance

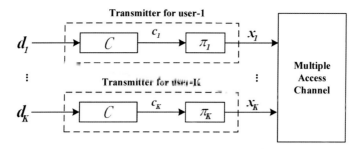

Fig. 2.2 The transmission structure of IDMA with K multiplexed users

the performance of asynchronous code division multiple access (CDMA). It has the benefits of preventing the effect of fading and mitigating the ICI as in CDMA [48]. Researchers in [49] expound that IDMA can exhibit some other attractive characteristics such as flexible rate adaptation, frequency diversity and power efficiency. Besides, the theoretical study of IDMA also shows that the interleaved low-rate codes with a simple chip-by-chip iterative decoding strategy could achieve the capacity of a Gaussian MAC [50].

The transmission structure of IDMA is illustrated in Fig. 2.2. The low-rate FEC encoder C is applied to encode the user-k's data bits d_k. The output is referred as coded bits c_k. The coded bits c_k pass through the interleaver π_k, after which multiple users' signals are multiplexed in the air. The interleaving patterns are generated independently and randomly, and vary from each other in order to distinguish the users. Therefore, the design of reasonable interleavers is rather essential. A user-specific interleaver design method is proposed in [51], which can resolve the memory cost problem and reduce the signaling exchanging between the gNB and the users. Besides, to accomodate IDMA in multi-carrier tranmission, e.g. in OFDM, a multicarrier interleave-division-multiplexing-aided IDMA (MC-IDM-IDMA) is prensented in [52].

IDMA has been wildly studied because of its robustness and user overload tolerance [19]. The structure of IDMA in single-path and multi-path environments is elaborated in [53]. Besides, a power allocation method is introduced to enhance the performance of IDMA by taking the advantage of the semi-analytical technique [48].

2.2.1.3 IGMA

Interleave-grid multiple access (IGMA) goes one step further than IDMA by introducing the grid mapping patterns [20], which can cooperate with the interleaving patterns to distinguish the signal streams from different users.

The flexibility to choose bit-level interleavers and/or grid mapping pattern for distinguishing the users could be easily supported in IGMA. Meanwhile, the

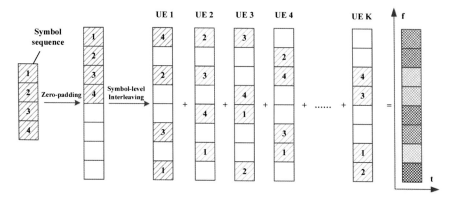

Fig. 2.3 An illustration of the grid mapping procedure of IGMA

scalability supporting different connection densities would be achieved with the abundant signatures generated by bit-level interleavers and grid mapping patterns.

Hereinafter, we briefly explain the general procedure of IGMA. Firstly, the user's data bits are encoded by the channel encoder to generate the coded bit sequence. The sequence is then interleaved to randomize the order of coded bits based on a pre-configured interleaver. The interleaved bit sequence is then modulated into the symbol sequence. Finally, the grid mapping process is conducted to interleave the symbol sequence as shown in Fig. 2.3. The whole procedure of IGMA can further benefit in combating frequency selectivity and ICI due to the randomization.

2.2.1.4   LSSA

The low code rate and signature based shared access (LSSA) scheme is proposed to support asynchronous massive transmission in uplink [23]. LSSA randomizes the MUI among the users by multiplexing the users' data streams with user-specific signature patterns at bit-level, where the signature patterns are usually unknown to others. Besides, the channel coding scheme which has very low code rate is adopted to encode each user's information bits in LSSA, which helps to mitigate the effect of the MUI. The low rate FEC code can also be replaced by employing higher rate FEC code along with spreading. After the channel coding, bit-level multiplexing with user-specific signature would be used. The user-specific signature may relate to the reference signal, the complex/binary sequence, and the permutation pattern of a short length vector. The length of orthogonal spreading codes is a factor that influences the number of simultaneous transmissions. Fortunately, the receiver in LSSA does not depend on orthogonal multiplexing codes to distinguish the target users' signals. Instead, the interference cancellation is exploited, so that high user overloading is well supported. The signature of LSSA can be chosen randomly at the user side or assigned to the user by the gNB. Furthermore, LSSA can also be optionally modified to have a multi-carrier variant in order to exploit frequency diversity provided by wider bandwidth and to achieve lower latency.

2.2.1.5   LCRS

Low code rate spreading (LCRS) is another NOMA scheme which utilizes the bit-level repetition and low rate coding to spread information bits over the total non-orthogonal transmission area [21]. Therefore, LCRS can achieve the maximum coding gain by combining channel coding and spreading through low rate codes. Under this circumstance, a user-specific channel interleaver [48] can be further exploited to aid the multi-user signal separation at the receiver.

2.2.2   Symbol-Level Non-orthogonal Multiple Access

Different from bit-level NOMA schemes which focus on the *bits*, symbol-level NOMA schemes play with *symbols* and mainly lay emphasis on the bit-to-symbol mapping. As illustrated in Table 2.2, a large portion of symbol-level NOMA schemes utilize the short sequence-based spreading to enhance the connectivity. These schemes can be further divided into two sub-categories according to the densities of the spreading sequences. Some other symbol-level NOMA schemes make use of long sequence-based scrambling/spreading/permutation, where the receiver exploits the difference between these sequences. Table 2.3 compares the pros and cons of applying long or short sequences in symbol-level NOMA, which are further illustrated in the following subsections.

Short Sparse Spreading NOMA

2.2.2.1   SCMA

Sparse code multiple access (SCMA) is a low density spreading-based NOMA scheme, which can achieve high overloading while maintain high reliability [32, 54, 55]. The core idea of SCMA is to directly map the coded bits to the multi-dimensional modulation symbols, according to a pre-defined sparse codebook, instead of sequentially conducting modulation and low density spreading. Therefore, both the resource element mapping and the multi-dimensional constellation are essential designs in SCMA [56]. The transmission process of SCMA is illustrated in Fig. 2.4, where multiple signal layers are multiplexed on the same radio resources. One major design aspect of SCMA is the sparse multi-dimensional codebook [57]. In [57], a SCMA codebook design method based on rotation, shuffling, and permutation is proposed, along with an example SCMA codebook which brings large shaping gain. SCMA

Table 2.3 Distinction between long and short sequences

	Long sequence	Short sequence
Level of operation	Bit level/symbol level	Usually symbol level
Generation of sequence	Randomly	Carefully designed
Usage	Disperse the encoded bit sequences so that the adjacent bits are approximately uncorrelated	To facilitate MUD
Receiving technique	Require iterative detection between symbol-level and bit-level, e.g. ESE-SIC	Symbol level detection, e.g. MPA/EPA
Synchronization requirement	Support asynchronous transmission when combined with single carrier waveform, e.g. RSMA	Synchronization is usually required, e.g., SCMA
Blind detection	Not support	Support

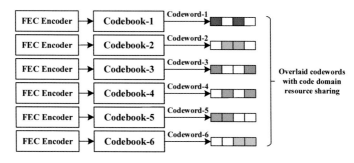

Fig. 2.4 A typical transmission structure of SCMA with six users and four subcarriers

can also achieve the signal space diversity by permuting the signal components to paired symbols located in multiple radio resources. Besides, with the sparse structure of SCMA, iterative multi-user detection algorithms, e.g. message passing algorithm (MPA), can be applied to simultaneously detect multiple data streams in symbol-level.

However, one concern of SCMA is that the sparse structure may be violated when single carrier is performed [23]. And MPA receiver may cause large computational burden and processing delay when the number of multiplexed users is large. Hence, a good tradeoff ought to be achieved between complexity and performance in the design of SCMA.

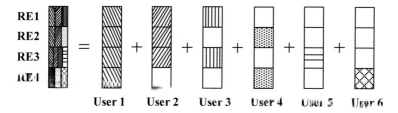

RE1
RE2
RE3
RE4

User 1   User 2   User 3   User 4   User 5   User 6

Fig. 2.5 Resource mapping of PDMA, with six users and four subcarriers

2.2.2.2 PDMA

Inspired by unequal transmission diversity and sparse coding, pattern division multiple access (PDMA) is proposed as a novel NOMA scheme to enhance the performance of multi-user communication system [28]. Different from SCMA which utilizes regular spreading signatures, PDMA usually employs irregular sparse signatures to facilitate the SIC receiving [58]. Besides, with the irregular sparse spreading signatures, PDMA can have a total number of 2^N signatures where N is the length of spreading.

An example of the code domain pattern matrix of PDMA is shown in (2.1), which involves six users and four subcarriers. A '1' means that the subcarrier is occupied by a user. According to the spreading patterns, the signals of the six users are illustrated in Fig. 2.5. However, we also see that one drawback of PDMA is that it cannot guarantee to accommodate the strict sparsity constraints as in SCMA, i.e., four users multiplex on the 3rd RE.

$$G_{\text{PDMA}} = \begin{bmatrix} 1\ 1\ 1\ 0\ 0\ 0 \\ 1\ 1\ 0\ 1\ 0\ 0 \\ 1\ 1\ 1\ 0\ 1\ 0 \\ 1\ 0\ 0\ 1\ 0\ 1 \end{bmatrix} \tag{2.1}$$

To further enhance the ability in distinguishing the multiplexed users, PDMA allows to utilize multiple domains in the design of the signature matrix, including temporal, spatial, code, power, and interleave domains [59, 60]. For example, PDMA with large-scale antenna array (LSA-PDMA) is proposed where the spreading signatures are designed jointly in beam and power domain to improve the system sum rate and access connectivity, respectively [61]. In addition, an interleaver-based PDMA (IPDMA) scheme is proposed, where the signal separation can be done according to different bit-level interleavers and/or characteristic patterns [62]. With the joint design at the transmitter and the receiver, PDMA can meet the need of higher spectral efficiency in 5G, while ensuring a reasonable receiver complexity.

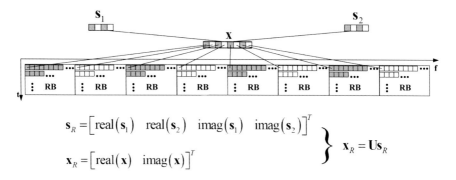

$$\mathbf{s}_R = \begin{bmatrix} \mathrm{real}(\mathbf{s}_1) & \mathrm{real}(\mathbf{s}_2) & \mathrm{imag}(\mathbf{s}_1) & \mathrm{imag}(\mathbf{s}_2) \end{bmatrix}^T$$
$$\mathbf{x}_R = \begin{bmatrix} \mathrm{real}(\mathbf{x}) & \mathrm{imag}(\mathbf{x}) \end{bmatrix}^T \qquad \Big\} \quad \mathbf{x}_R = \mathbf{U}\mathbf{s}_R$$

Fig. 2.6 An illustration of LDS-SVE

2.2.2.3 LDS-SVE

Low density signature-signature vector extension (LDS-SVE) is another LDS-based NOMA scheme [22]. The major difference between LDS-SVE and the other LDS-based NOMAs, i.e., SCMA and PDMA, is that the former introduces user-specific signature vector extension, which is performed by transforming and concatenating two element signature vectors into a larger signature vector.

In the following, we show an example of LDS-SVE in Fig. 2.6. The modulated symbols are first divided into two vectors, i.e., $\mathbf{s}_i$, $i = 1, 2$, according to a serial-to-parallel transformation. Define $\mathbf{s}_R$ as a real vector obtained by stacking the real and imaginary parts of signature $\mathbf{s}_1$ and $\mathbf{s}_2$, as shown in Fig. 2.6. The SVE output is defined as a real vector $\mathbf{x}_R$, which is achieved by multiplying $\mathbf{s}_R$ with a transformation matrix $\mathbf{U}$. At last, transmission complex signal $\mathbf{x}$ can be recovered from $\mathbf{x}_R$, i.e., by reconstructing the complex symbols from the real vector $\mathbf{x}_R$. The main advantage of LDS-SVE is that, by multiplying $\mathbf{U}$, the original modulation symbols are spread on more REs, which brings higher order of diversity.

Short Dense Spreading NOMA

2.2.2.4 MUSA

Multi-user sharing access (MUSA) is a NOMA scheme based on short complex spreading sequence and SIC receiver [24]. In general, the spreading sequences in MUSA does not have sparsity as in SCMA, PDMA, and LDS-SVE [14]. We illustrate the transmission procedure of MUSA in Fig. 2.7. After channel encoding and modulation, as shown in Fig. 2.7, each user's data symbols are spread by a complex sequence, whose elements take values in complex field. Then the spread symbols of each user are transmitted on the shared radio resources. At the receiver, the well-designed spreading sequences are exploited by the multi-user detectors to distinguish

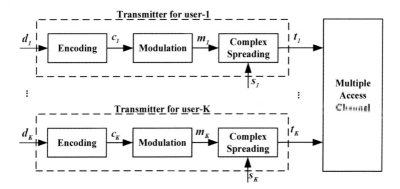

Fig. 2.7 The transmission structure of MUSA with K simultaneous users

different users' data streams. It is noteworthy to mention that different symbols of the same user may use different spreading sequences, which can average the MUI and improve the system-level performance.

Short sequence-based spreading is the major operation in the MUSA transmitter. Each user can randomly pick one spreading sequence from a sequence pool consisting of multiple spreading sequences. The spreading sequence design in MUSA follows the guidelines of low cross-correlation, where each element of the sequence is chosen out of a complex scalar set, e.g. $\{\pm 1, 0, \pm i\}$. Due to the utilization of the imaginary part, complex spreading sequences could perform the lower cross-correlation compared to pseudo random noise (PN), even with a short spreading length [63]. In addition, with arbitrarily selected complex elements, the pool of the spreading sequences in MUSA can be very large.

2.2.2.5 NCMA

Similar to MUSA, non-orthogonal coded multiple access (NCMA) also uses non-orthogonal dense spreading sequence to minimize MUI and support high overloading capability [25]. The spreading sequences of NCMA, also named as non-orthogonal cover codes (NCC), is obtained by solving the Granssmannian line packing problem, where the solutions of the problem guarantee the optima non-orthogonal sequences [64]. Due to the design of NCC, the interference level between two users is predictable.

The transmission structure of NCMA is illustrated in Fig. 2.8. In NCMA, each user's data symbol is spread with NCC, and an additional FFT operation can be implemented before IFFT to reduce the PAPR. At the receiver, a simple de-spreading and parallel interference cancellation (PIC) detector can be implemented to recover the multiplexed signals. We note that applying IFFT on spare spreading-based NOMA schemes, e.g. SCMA and PDMA, would destroy the sparse structure and lead to high computational complexity in detection. To further improve the connectivity and bring

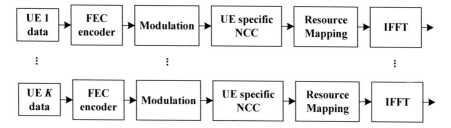

Fig. 2.8 The transmission structure of NCMA with K simultaneous users

additional throughput gain under specific QoS constraints, multi-stage spreading based on NCC can be applied. However, the correlation properties of the multi-stage spreading sequences, which are composed by multiplying several NCCs, need to be clarified. Hence, a good trade-off between the connection density and the decoding performance needs to be further evaluated [23].

2.2.2.6 NOCA

Non-orthogonal coded access (NOCA) is another spreading-based NOMA scheme. Similar to other symbol-level NOMA schemes, the data symbols in NOCA are spread according to non-orthogonal sequences before transmission [26]. The spreading in NOCA is operated in both time and frequency domain. We demonstrate the transmission structure of NOCA in Fig. 2.9. The serial modulated symbol sequence is first converted to P parallel subsequences by a S/P converter. We denote C_j as the non-orthogonal spreading sequence with length SF, where SF denotes the spreading factor. The j-th subsequence is then spread on SF subcarriers according to C_j. Hence, a total number of $P \times SF$ subcarriers are required for NOCA. Besides, to accommodate the single-carrier transmission in uplink, FFT operation can also be applied before IFFT to reduce the PAPR.

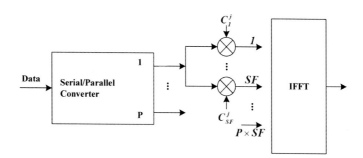

Fig. 2.9 The transmission structure of NOCA

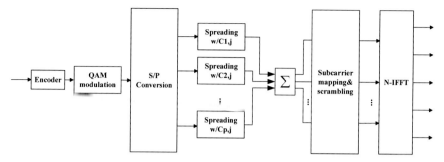

Fig. 2.10 The transmission structures of SSMA

To ensure high detection accuracy and high overloading, the spreading sequences used in NOCA should follow some properties, such as good auto-correlation, low cross-correlation, and low storage requirement. Meanwhile, the sequences should have constant modulus to ensure low cubic metric. Besides, multiple spreading factors might be supported for flexible adaptation.

2.2.2.7 SSMA

Short sequence spreading-based multiple access (SSMA) is another spreading-based NOMA scheme [21], which directly spreads the modulation symbols with multiple orthogonal or quasi-orthogonal codes, and transmits the spread symbols in time-frequency resources allocated for non-orthogonal transmission. The transmission structure of SSMA, as illustrated in Fig. 2.10, is similar to NOCA, where user-specific scrambling is applied to average the MUI.

Long Sequence-Based NOMA

2.2.2.8 RSMA

Resource spread multiple access (RSMA) is a novel NOMA scheme which applies long spreading or scrambling sequence to disperse the users signal over the entire radio resources. In RSMA, each user's codewords can be spread over all available time and frequency resources [24]. Therefore, RSMA can achieve full diversity compared to short spreading-based NOMA schemes. At the receiver, different spreading/scrambling sequences can be exploited to distinguish different signal streams. Besides, low rate FEC codes and advanced detection algorithms in RSMA can ensure high transmission reliability. The scrambler can also be replaced by different interleavers for the sake of whitening the MUI.

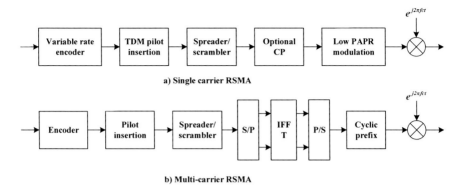

Fig. 2.11 The transmission structures of singl-carrier and multi-carrier RSMAs

According to different application scenarios, two kinds of RSMA schemes have been proposed, i.e., single-carrier RSMA and multi-carrier RSMA [30], as shown in Fig. 2.11. On the one hand, single-carrier RSMA employs the single-carrier waveforms and low PAPR modulations to enhance the performance of battery power consumption and coverage extension for small data transmission. Match-filter (MF) based receiver can be applied to distinguish different signals of single-carrier RSMA with low computational complexity. In addition, single-carrier RSMA does not rely on joint detection, which looses the synchronization requirement and makes it a good candidate for asynchronous access. On the other hand, multi-carrier RSMA is studied to lower the latency and to promote the spectral efficiency for legacy users.

2.2.2.9 RDMA

Repetition division multiple access (RDMA) can be regarded as an interleave-based NOMA scheme [29]. However, instead of deploying bit-level interleaving as in IDMA, RDMA focuses on the symbol-level interleaving which is designed based on simple cyclic-shift repetitions. In RDMA, each user's modulation symbol vector is repeatedly transmitted for several times, where different cyclic shift indexes are assigned to the repetitions. Besides, different users would have different repetition and cyclic-shift patterns, which enables completely randomized MUI and achieves both time and frequency diversities.

The transmission structure of RDMA with K simultaneous users is illustrated in Fig. 2.12. Compared with IDMA and RSMA, RDMA is simpler and may reduce the signaling overhead, since the user-specific scrambling and interleaving patterns are not needed. Meanwhile, SIC receiver is used in RDMA to provide good trade-off between receiving complexity and detection performance.

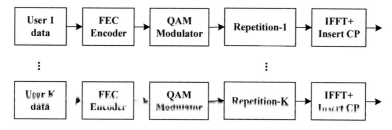

Fig. 2.12 The transmission structure of RDMA with K simultaneous users

2.2.2.10 GOCA

Group orthogonal coded access (GOCA) is another long sequence-based NOMA scheme, which can be seen as an enhanced version of RDMA [29]. The major difference between GOCA and RDMA lays in that the former employs the group orthogonal sequences to spread the modulation symbols into shared time and frequency resources after repetitions, as shown in Fig. 2.13. Similar to RDMA, SIC receiver is expected to achieve good detection performance with moderate computational complexity.

The group orthogonal sequences have a two-stage structure, where orthogonal sequences and non-orthogonal sequences are used in first and second stage, respectively. Therefore, as shown in Fig. 2.14, we can divide the GOCA sequences into several non-orthogonal groups according to the non-orthogonal sequences used in the second stage, while the sequences within a group are orthogonal to each other due to the design in the first stage.

2.2.3 Multi-user Detection Technologies

According to NOMA protocol, different users' signal streams are multiplexed on the same radio resources, therefore the multi-user detection (MUD) technologies are needed to distinguish independent signal streams. In the sequel, we analyze some

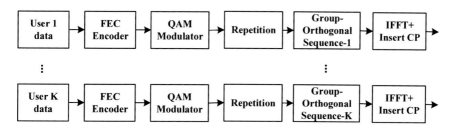

Fig. 2.13 The transmission structure of GOCA with K simultaneous users

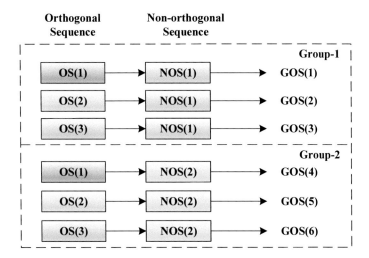

Fig. 2.14   An example of group orthogonal sequences in GOCA, with two groups and a total of six sequences

essential MUD technologies, which are proposed to match the NOMA schemes in the above subsections.

2.2.3.1   MMSE-SIC

The minimum mean square error-successive interference cancellation (MMSE-SIC) receiver is a direct extension of MMSE receiver, as shown in Fig. 2.15. In the first iteration of MMSE-SIC, the signal with largest received SINR is first detected by MMSE receiver by regarding the interference as noise, demapped and then decoded to obtain the information bits. After that, the signal of this user is reconstructed and canceled from the received signal. The above procedure is repeated in the following iterations until no signal stream can be successfully recovered.

MMSE-SIC receiver suffers from error propagation problem, where the estimation errors in previous signal layers may propagate to the remaining layers. With the aim to mitigate the error propagation, the received SNRs of different data streams shall have large differences to ensure sufficient SINR in each iteration. Therefore, MMSE-SIC is especially suitable for PD-NOMA, as well as other NOMA schemes where users have diversified channel conditions.

2.2.3.2   MPA

MPA is an iteration-based non-linear symbol detection algorithm, which can exploit the structure of sparse spreading sequences and achieve near maximum-likelihood

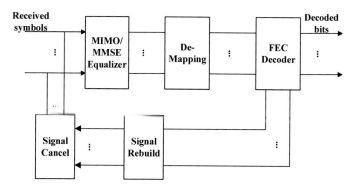

Fig. 2.15 The structure of MMSE-SIC receiver

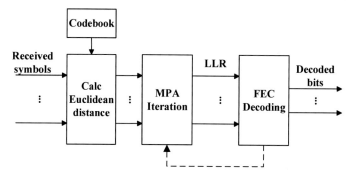

Fig. 2.16 The structure of MPA receiver .eps

(ML) performance. Different from ML receiving which estimates the entire spreading block with full search method, MPA only conducts localized optimal detection on each resource element to acquire the soft information about the transmitted symbols, and then delivers the information to the neighboring resource elements as the extrinsic information of the next localized optimal detection. We show the MPA receiver in Fig. 2.16.

The original MPA receiver focuses on symbol-level detection, and does not exploit the error correction ability of FECs to separate different signal streams. To resolve this problem, MPA-turbo and MPA-SIC are proposed as shown in Figs. 2.17 and 2.18,

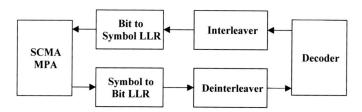

Fig. 2.17 MPA-turbo

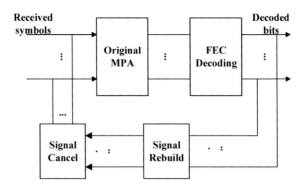

Fig. 2.18 MPA-SIC

respectively. As revealed in its name, MPA-turbo works just like the Turbo decoding, where the FEC decoder processes the soft information provided by MPA and then feedbacks the processed soft information to MPA module as extrinsic information. Different from MPA-turbo, MPA-SIC directly cancels the recovered signal streams from the received signal to mitigate the MUI.

2.2.3.3 EPA

Although MPA significantly reduces the computational complexity compared to ML, the complexity still grows exponentially with the number of multiplexed users on each radio resource. Estimation propagation algorithm (EPA) is another graphical-based multiuser detection algorithm, proposed for SCMA, to further reduced the computational complexity order from exponential to linear [65]. The idea of EPA originates from the variational approximate inference method, which is commonly applied in the machine learning era [45]. Different from MPA, EPA employs a Kullback-Leibler divergence based projection in the message update steps to align with the expectation propagation principle. We can directly replace the MPA module with the EPA module in the SCMA receiver, as shown in Fig. 2.19, and generate new variants of EPA such as EPA-turbo and EPA-SIC using the similar approaches in MPA.

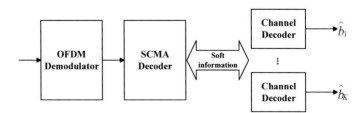

Fig. 2.19 EPA

2.2.3.4 ESE-PIC

Elementary signal estimation-parallel interference cancellation (ESE-PIC) receiver is originally proposed in IDMA, which has shown robust performance even when a large number of users are multiplexed. As shown in Fig. 2.20, ESE-PIC first detects transmitted symbols via ESE detection, a linear symbol detector. Then the detected signals are parallelly deinterleaved and decoded to acquire the coding gain. The output information from the decoder is then sent back to ESE module to aid symbol detection.

2.2.3.5 Comparisons on MUD Receivers

We now compare the pros and cons of the aforementioned receiving technologies, as well as their applicable NOMA schemes, as shown in Table 2.4. To sum up, the overall structure of MUD receivers constitutes of two parts, i.e., symbol detector and FEC decoder. Joint symbol detection, i.e., MPA and EPA, achieves better performance than single user detection, i.e., MMSE. However, they only work with short and sparse spreading sequences. Long sequence based schemes require iterative detection, i.e., ESE-PIC, however, due to parallel message passing, several decoders may work at the same time which leads to even larger hardware cost than MPA-turbo/SIC. To facilitate the implementation of NOMA and satisfy the diversified requirements of 5G, a good tradeoff between detection accuracy, computational complexity, latency, and hardware requirements should be achieved, which certainly requires further study.

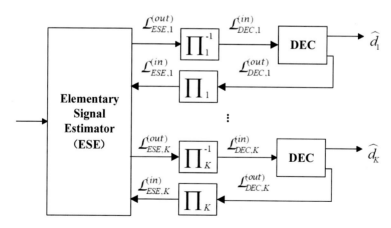

Fig. 2.20 ESE-PIC

Table 2.4 Comparisons of NOMA receivers

Receiver	Main character	Pros	Cons	Applications
MMSE-SIC	Reuse single-user receiver and SIC	Low complexity	Require large SNR gaps among users	Almost all schemes, especially PD-NOMA
MPA/EPA	Symbol-level iterative detection	Near-ML symbol detection	Middle/high complexity	Sparse spreading NOMA
MPA-turbo/SIC	Iterative detection between symbol-level and bit-level	Better performance than MPA/EPA	High complexity	Sparse spreading NOMA
ESE-PIC	Iterative detection between symbol-level and bit-level	No requirement on sparsity	High complexity & hardware overhead	Especially bit-level NOMA

2.3 Grant-Free Multiple Access for mMTC

The state-of-art NOMA schemes, mentioned in Sect. 2.2, are mainly based on centralized scheduling, where spreading sequences, interleaving patterns and/or transmission powers of different users are scheduled by the gNB. However, the major drawback of the scheduling-based NOMA is that the signaling overhead occupies a large portion of radio resources, which makes the grant-free NOMA inevitable. In the following section, we analyze the motivation and the procedures of grant-free NOMA, as well as the typical transmission and reception technologies.

2.3.1 Motivation

The conventional human-type communications are normally optimized for mobile broadband (MBB) services [66], with small amount of users, high data rate, large packet size. Compared with the size of data packet in MBB, the control signaling is relatively few. Therefore, the human-type communications are not sensitive to the signaling overhead, and usually involve frequent interactions between the gNB and the users to maintain high reliability and high data rate.

Despite the MBB services, 5G also aims at supporting mMTC, where massive connectivity and long battery life cycle are two key requirements. Different from in MBB scenario, the arrival of data packets in mMTC is sporadic and the packet sizes are rather small [67]. Based on [1], mMTC would support more than a million devices per square-kilometer, and this, together with the small packet sizes, make the control signaling overhead rather significant. Therefore, the simplification on access procedure, as well as the reduction on signaling overhead are both needed to satisfy the massive connectivity requirement of mMTC.

Grant-free NOMA, where multiple users conduct uplink instant transmissions without grant, can significantly reduce the signaling overhead. It is agreed that grant-free NOMA is more suitable for mMTC scenario due to the following concerns [66]:

• Energy saving: Resource allocation has been well studied to extend the battery times of the devices, however, with a waste of the signalling overhead. Grant-free access can save the energy of the devices, i.e., the devices can decide to turn to active mode when small packets arrive, or keep in sleep state if transmission is not needed;
• Low cost devices: grant-free access can trade the computational complexity at the gNB with the hardware cost at the devices;
• Latency and signaling overhead reduction: no additional latency or signaling overhead is induced by the signaling interactions.

Out of the above considerations, 3GPP has agreed that NR should target to support uplink "autonomous/grant-free/contention based" at least for mMTC scenario [68].

2.3.2 Grant-Free Process

In Fig. 2.21, we compare the signaling procedures between the scheduling-based transmission in LTE and the grant-free transmission. In LTE, when a user becomes active, it would conduct random access procedure firstly, which includes at least 4 steps, i.e., preamble transmission, random access response (RAR), Message. 3 and Message. 4. After that, a scheduling request (SR) is transmitted to the gNB if the buffer is not empty, and the user does not transmit packets until it receives the uplink grant from the gNB. The above mentioned procedures may take dozens of millisecond, which impose large signaling overhead on the network, consume more power for the signaling transmission/detection on the device side, and incur large latency for the data transmission. In contrast, grant-free transmission achieves autonomous transmission without explicit dynamic grant. Compared with the scheduling-based transmission, grant-free reduces signaling overhead, as well as control/user plane latency [69]. We note that, due to the decentralized uplink instant transmissions, the signals of users are multiplexed, which naturally leads to non-orthogonal transmissions.

We show the state graph of a grant-free user in Fig. 2.22. If the user has no data in buffer, it stays in a sleep state, otherwise, it would wake up, synchronize according to reference signals and acquire some necessary system broadcast information and some pre-defined uplink grant information. Before directly transmitting information block with the grant-free manner, preamble may be transmitted for the uplink synchronization for detection in the receiver side. Furthermore, some multiple access information could be implicitly indicated by the preamble, such as spreading signature, locations of radio resources as well as the timing of retransmission, etc. With

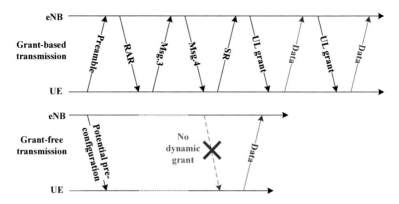

Fig. 2.21 Comparison between the general procedures of grant-based and grant-free

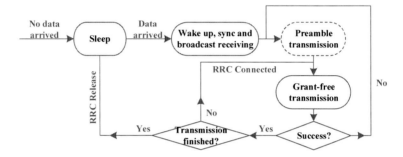

Fig. 2.22 Grant-free uplink transmission illustrations

these information, the collisions can be detected, and the blind detection complexity of the gNB can also be greatly reduced.

According to whether random access channel (RACH) is required, grant-free transmission can be classified into RACH-based grant-free and RACH-less grant-free.

2.3.2.1   RACH-Based Grant-Free

When all the users have performed RACH, grant-free transmission would occur in a more synchronized manner, i.e., the timing offsets among users are mostly within the cyclic-prefix (CP) length. Therefore, this type of grant-free transmission is referred as RACH-based grant-free [70] since RACH procedures have been done before data transmission. RACH-based grant-free could also reduce the overhead of SR and uplink grant, and at the same time, it is beneficial for signal detection.

2.3.2.2 RACH-Less Grant-Free

In order to reduce the signaling overhead, the RACH procedure could also be canceled, i.e., data transmission phase starts whenever there are packets arriving. This method can be referred to as RACH-less grant-free [70, 71]. In this way, not only the RACH associated signaling, but also the battery energy can be saved, since the user can go to sleep if there is no data to transmit. However, the absence of RACH may result in the asynchronization among users, which may cause large detection complexity at the receiver.

2.3.3 Typical Grant-Free Multiple Access Technologies

In this subsection, we analyze the typical grant-free NOMA technologies, which are categorized into two classes, i.e., grant-free bit/symbol-level NOMA schemes, and graph-based access, according to different design principles.

2.3.3.1 Grant-Free Bit/Symbol-Level NOMA

Grant-free bit/symbol-level NOMA can be obtained by directly incorporating grant-free access protocol with the existing NOMA schemes mentioned in Sect. 2.2, especially the short-spreading based NOMA, e.g. SCMA, PDMA and MUSA. To enable grant-free access in NOMA, a contention-based unit (CTU) is defined as the basic multiple access resource, as shown in Fig. 2.23, where each CTU may constitute of several fields including radio resources, reference signal, and spreading sequence, etc [55]. One CTU may differ from the others in any fields, and these differences can be exploited by the receiver to distinguish different signal streams.

When a user has data in buffer, it randomly selects a CTU and then transmits its data packet accordingly, i.e., spreading the modulation symbols with the given spreading sequence over the given radio resources, as well as the given reference

Fig. 2.23 An illustration of CTU

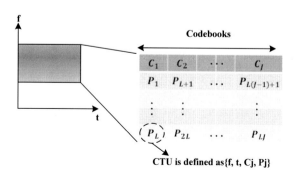

signals. When the number of active users is small, the users may choose different radio resources, and hence they are orthogonal. Otherwise, when the number of active users is large, the signals of the users are overlapped on the radio resources, and other fields in the CTU come into play in MUD. Therefore, grant-free NOMA can be regarded as a generalized orthogonal and non-orthogonal access.

The spreading sequences in CTUs can reuse the sequences designed for SCMA, PDMA or MUSA. Spreading can also be replaced with sparse repetition, as proposed in [72], where simple inter- and intra-slot SIC can be employed to recover multiple signal streams. In the meantime, with sparse spreading sequences, the MUI is mitigated and the receiving complexity also remains low. On the other hand, with dense spreading sequences, more diversity gain can be achieved which may combat the fading of wireless channel.

Due to the uncoordinated transmissions, the collision among users would be a severe problem in grant-free symbol-level NOMA. A hard collision happens when several users choose the same CTU. Under such circumstances, these users may be distinguished and detected only if they have distinctive channel gains. From this perspective, it is important to enlarge the pool of the spreading sequences, as have done in MUSA [67]. However, enlarging the pool size may also increase the cross-correlations among the sequences, which may degrade the transmission reliability. Therefore, a good tradeoff between the pool size and the cross-correlations should be achieved to mitigate the collisions while maintain high reliability.

2.3.3.2 Graph-Based Access Schemes

Slotted ALOHA (SA) is a conventional uncoordinated random access method which was proposed in 1970s. Recently, a class of graphical-based random access schemes has been proposed which introduces the theory of linear coding into ALOHA access [73–82]. The main idea of these schemes is to regard the random access as random coding, and to optimize the access probability with coding theory. Inter-slot SIC receiving is applied to deal with the collisions. Besides, density evolution (DE) algorithm is usually applied to design or evaluate the transmission patterns of these schemes.

In [75], contention resolution diversity slotted ALOHA (CRDSA) is proposed which combines repetition codes with SA where two replicas of each burst is transmitted randomly in two slots, and the collided received bursts are divided by the SIC algorithm. An enhanced scheme of CRDSA, named irregular repetition slotted ALOHA (IRSA), is also introduced in [75] which optimizes the transmission method in CRDSA by bipartite graph and allows more feasible repetition pattern than CRDSA. In [73], a more generic schemes proposed, named coded slotted ALOHA (CSA), which encodes the bursts via linear block code instead of the replicas, and combines the iterative SIC with linear block code decoding to recover the source packets. A frameless ALOHA scheme based on rateless codes is provided in [83] where the transmissions of bursts act as the encoding process of rateless codes. Then, the receiver would send a feedback to the transmitter when its burst is recovered.

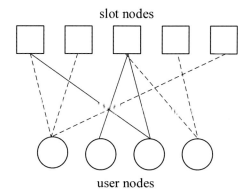

Fig. 2.24 Bipartite graph representation of CSA

slot nodes

user nodes

Hereinafter, we show a bipartite graph representation of the transmissions of CSA in Fig. 2.24, where 4 users transmit bursts within 5 slots. Each burst node denotes the burst belonging to a user, each slot node denotes a slot and each edge denotes that the replica of the corresponding burst is transmitted in the corresponding slot. Meanwhile, we also show the SIC process by a bipartite graph in Fig. 2.25. In each iteration of SIC, the bursts occurred in the slots without collisions can be recovered immediately, thus the edges connected to these bursts can be removed. Then the next iteration starts and the iterations continue until no slots can be recovered. The nodes in green denote that the bursts of these users have been already recovered in the previous iterations.

2.3.4   Detection Techniques

In grant-free NOMA systems, the users randomly select the resources to transmit data without the dynamic scheduling. Therefore, the aforementioned MUD technologies in Sect. 2.2.3, where the identities of active users and their selected signatures are known to the receiver, is unreasonable in the practical grant-free NOMA system. Blind detection, where user activation, channel coefficients, and data packets are simultaneously detected, should be studied. Furthermore, since the transmission phase of grant-free NOMA is pretty simple, the complexity is transferred to the receiver side, which makes it rather important to design efficient blind detection algorithms.

We illustrate the general procedure at the grant-free NOMA receiver in Fig. 2.26. The whole receiving process can be divided into two stages, i.e., user activity activation stage and data detection stage. In the first stage, active users are identified out of a potential user list. Then, in the second stage, the channel coefficients are estimated and the data packets are detected.

The idea of compressive sensing (CS) can be incorporated into the first stage due to the fact that the user activation is sparse. This sparsity is utilized in the CS-MPA

Fig. 2.25 Bipartite graph
representation of SIC process

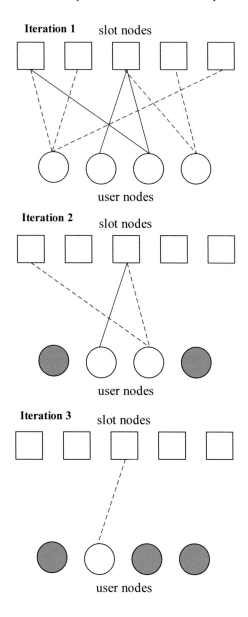

Fig. 2.26 Grant-free
NOMA receiver

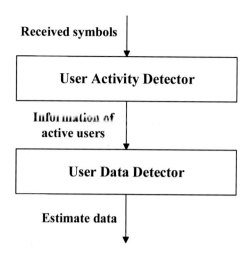

detector which jointly uses CS and MPA to realize both stages simultaneously [84]. Compared with the conventional MPA without activity detection, it achieves better BLER and throughput. In addition to CS, the user activity detection could be realized by different algorithms and schemes. For example, focal under-determined system solver (FOCUSS) and expectation maximization (EM) are proposed and analyzed for active pilot detection [85], and they can be combined with the blind data detection method, i.e., joint data and active codebook detection (JMPA), to recover the data in the spreading-based grant-free systems. It is seen that JMPA can achieve scarcely any performance degradation in decoding users' data without prior knowledge of active codebooks. Furthermore, to avoid the redundant pilot overhead, a novel sparsity-inspired sphere decoding (SI-SD) algorithm is proposed by introducing one additional all-zero codeword to achieve the maximum a posteriori (MAP) detection [86]. However, either CS or EM can only get the rough information about the active users. Detection-based group orthogonal matching pursuit (DGOMP) is a user activation detector which is promising to get a more accurate active user set [87]. Meanwhile, an enhanced version of JMPA is proposed in [87], which takes the channel gain and noise power into consideration when calculating the prior information of the zero codeword. The modified JMPA also helps to eliminate the false detections caused by noise, channel fading and non-orthogonality of pilot sequences.

2.4 Implementation Issues

Non-orthogonal transmission is completely different from the orthogonal transmission which has been widely implemented in LTE. As a consequence, non-orthogonal transmission arises some implementation issues for practical deployment. In this section, we analyze some important implementation issues related to scheduling-based NOMA schemes and grant-free NOMA, respectively.

2.4.1   Scheduling-Based Multiple Access

Recall that, the major difference between OMA and scheduling-based NOMA is that the latter allows multiple superimposed transmissions on the same radio resources, while the former only allows orthogonal transmissions. Hence, the resource allocation and demodulation reference signal (DM-RS) should be designed to facilitate scheduling-based NOMA.

2.4.1.1   Resource Allocation and Scheduling

Resource allocation, where the radio resources are assigned among users via centralized scheduling to meet certain optimization targets, has been extensively exploited in LTE to promote the system-level performance, including peak transmission data rate, average throughput, and user fairness. When multiple scheduling requests are transmitted from the users in LTE, the network orthogonally allocates the limited radio resources to a subset of the candidate users. However, the resource allocation in NOMA would be complex, since the resources in NOMA not only consists of radio resources, but also MA signature resources. Due to the fact that radio resources can be shared among users, the resource allocation problem would be even more complex. Besides, specific resource allocation methods should be designed for different NOMA schemes to match their unique characteristics. For example, PD-NOMA tends to multiplex the cell-center users and cell-edge users, while SCMA is more likely to superimpose the signals of co-located users.

The resource allocation in NOMA should also be designed to mitigate the effect of error propagation, if SIC-based receiver is employed. For example, the users with small SIC order should be allocated with more radio resources to transmit signals with lower coding rates. When multi-cell system is considered, resource allocation should be designed to reduce the ICI. As an instance, very-low density MA signatures may be allocated to the cell-edge users in uplink NOMA.

2.4.1.2   DM-RS

In LTE, DM-RSs with different cyclic shifts and orthogonal cover codes (OCCs) can be orthogonally multiplexed when the cyclic shift is longer than the channel delay spread. However, with the increasing demands for DM-RS ports in NOMA, it is unrealistic to add more cyclic shifts, and in the meantime, the OCC resources are also limited. Comb structures may be adopted in the design of DM-RS for NOMA. Figure 2.27 shows an example of the comb structure of DM-RS, which could increase

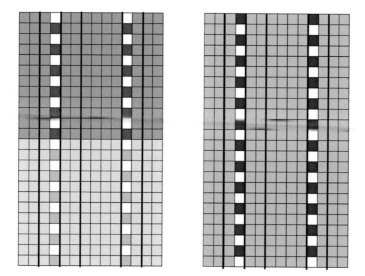

Fig. 2.27 An example of the DM-RS structures with different combs

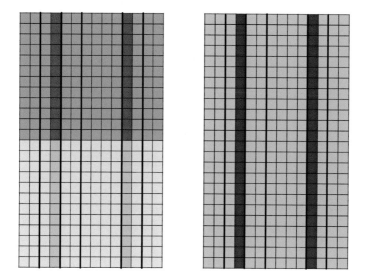

Fig. 2.28 An example of the DM-RS structure with different OCC

the number of DM-RS resources without decreasing the accuracy in channel estimation. Compared with previous DM-RS schemes, as shown in Fig. 2.28, the comb structure guarantees the orthogonal property of DM-RS via FDM [88, 89]. To support massive connectivity, non-orthogonal DM-RS may be further introduced to enlarge the number of DM-RS ports, where advanced channel estimation techniques should be exploited to mitigate the effect of non-orthogonality [90].

2.4.2   Grant-Free Multiple Access

As discussed in the previous section, data transmission in grant-free NOMA follows an arrive-and-go manner, which is very different from both OMA and scheduling-based NOMA. Therefore, implementing grant-free NOMA would require more efforts. This subsection presents several critical implementation issues related to grant-free NOMA, such as resource allocation, hybrid automatic repeat request (HARQ), link adaption, and physical signal design.

2.4.2.1   Resource Allocation

Similar to scheduling-based NOMA, the resources of grant-free NOMA also consist of radio resources and MA signatures. Two kinds of resource selection methods have been proposed in grant-free NOMA, i.e., random resource selection method and pre-configured method [91].

In random resource selection method, users randomly select the radio resources and the MA signatures, and then transmit signals accordingly. In this case, the user activities are not available at the gNB [92], which may cause the ambiguity. In order to resolve the above problem, the radio resources should be divided into orthogonal multiple access blocks (MABs) in the time and frequency domains, as shown in Fig. 2.29. Different MABs occupy different radio resources and may adopt different transmission settings, such as transmission block sizes (TBSs), modulation and coding schemes (MCSs), and transmission modes (TMs). The configurations of the MABs in a cell can be broadcasted by the gNB as system information. During data transmission phase, each active user first selects one MAB and then selects an MA signature. At the receiver, multi-user detection can be parallelly performed on each MAB. In addition, each MAB may be assigned with a limited number of MA signatures, which can reduce the computational complexity of blind detection at the receiver.

In the pre-configured resource allocation scheme, several users may be allocated the same radio resources along with unique MA signatures [93]. The MA signatures can be used for active user identification and collision reduction. We show an example

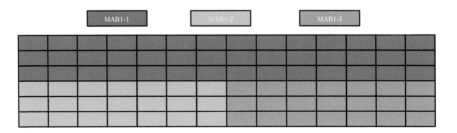

Fig. 2.29 An example configuration of MAB

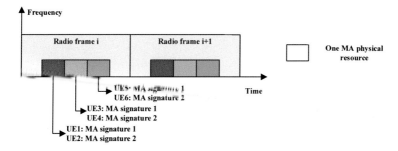

Fig. 2.30 An illustration of pre-determined MA resources for grant-free transmission

of pre-configured scheme in Fig. 2.30, where six users are scheduled with three distinctive pieces of MA radio resources, and the multiplexed users are allocated with different MA signatures.

2.4.2.2 HARQ

When the initial grant-free NOMA transmission is not successfully recovered, there is a need to retransmit the data for one or more times. HARQ is a profitable retransmission scheme which can merge the information of new transmission with previous transmissions in an effective way [94]. Due to the absence of uplink grant, one significant issue of supporting HARQ in uplink grant-free transmission is that, how does gNB identify the first transmission and the retransmissions for a HARQ process. One potential method is that the gNB can explicitly schedule retransmissions via downlink control signaling. Another method is to divide the MABs into several groups according to the maximum allowed number of retransmissions [95], where different users may select different MAB since they have different retransmission numbers. Another key issue in HARQ is the ACK/NACK indication. As discussed in [95], when collisions happen, gNB can utilize the RAR-style feedback, normally consisting of HARQ-ACK as well as user identification information, from which the collided users may identify whether or not their data are successfully decoded.

2.4.2.3 Link Adaption

The link adaptation has been introduced in LTE to adapt to the instantaneous channel condition by adjusting the transmission parameters. Properly design link adaptation not only results in low BLER, but also reduces the retransmission number and collision probability [96]. In addition, the suitable link adaptation could achieve lower latency, which is an important target in NR application scenarios [97].

In general, the link adaptation is realized by obtaining channel state information. However, the intermittent transmissions in uplink grant-free NOMA lead to the fact that the users might not be able to get accurate uplink channel status [98]. One solution is to use the measurements of downlink reference signals to determine the link adaption parameters for uplink transmission. The link adaption parameters may include MCS, number of repetitions, size of MA radio resources, and the transmission power during subsequent retransmissions, etc [99].

2.4.2.4 Physical Signal Design

Physical signals, including preamble and DM-RS, are another important design aspect in grant-free NOMA. Preamble has been used in LTE for random access request [100]. However, with the aim of reducing signaling overhead, the complete random access procedure may be omitted (e.g. RACH-less grant-free). Instead, the preambles are usually directly followed by data symbols, as shown in Fig. 2.31.

In RACH-less grant-free NOMA transmission, the users autonomously choose MA signatures as well as time instant for initial transmission, which are not known by gNB and may lead to asynchronization among received signals. In this case, well designed preambles could assist the active user identification, MCS indication, timing offset (TO)/frequency offset (FO) estimation, and channel estimation.

Similar to the preambles, DM-RS can be used for channel estimation and user identification in grant-free NOMA [89]. However, due to the uncoordinated transmissions, different users may choose the same DM-RS, which greatly degrades the accuracy of channel estimation. To guarantee low collision probability on DM-RS, sufficient number of orthogonal/semi-orthogonal DM-RSs should be provided [88]. Besides, advanced multi-user detection algorithms, such as SIC, could also help to increase the quality of channel estimation [27].

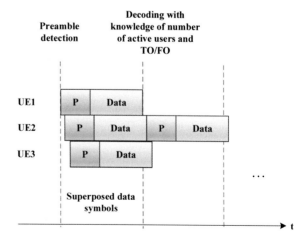

Fig. 2.31 An illustration of RACH-less grant-free NOMA transmissions, where the data symbols are transmitted immediately after preambles

2.5 Conclusions

NOMA has been recognized as one of the key enabling technologies to accomplish the diversified requirements of 5G. By enabling multiple users to share the same radio resources and exploiting the advanced MUD algorithms, NOMA exhibits better performance than OMA, especially in SE and connectivity. As demonstrated in this review, the idea of superimposing the users has been carried forward into multiple domains, including power, code, interleave, and scramble, which have motivated many NOMA schemes. Meanwhile, various multi-user receiving technologies also facilitate the application of NOMA in different scenarios. Besides, we also look into grant-free NOMA, which aims at reducing the signaling overhead and increasing the access probability for mMTC. We hope that our survey would shed a light on the deployment and development of NOMA technologies.

References

1. 3GPP TR 38.913, Study on scenarios and requirements for next generation access technologies
2. A.E. Gamal, Y.H. Kim, Lecture notes on network information theory. Mathematics (2010)
3. S. Shimamoto, Y. Onozato, Y. Teshigawara, Performance evaluation of power level division multiple access (PDMA) scheme, in *Proceedings of 1992 IEEE International Conference on Communications*, vol. 3 (1992), pp. 1333–1337
4. K.I. Pedersen, T.E. Kolding, I. Seskar, J.M. Holtzman, Practical implementation of successive interference cancellation in DS/CDMA systems, in *Proceedings of 1998 IEEE International Conference on Universal Personal Communications*m vol. 1 (1996), pp. 321–325
5. G. Mazzini, Power division multiple access, in *Proceedings of 1998 IEEE International Conference on Universal Personal Communications*, vol. 1 (1998), pp. 543–546
6. Y. Yan, A. Li, H. Kayama, Superimposed radio resource sharing for improving uplink spectrum efficiency, in *Proceedings of 2008 14th Asia-Pacific Conference on Communications* (2018), pp. 1–5
7. 3GPP TR 38.812, Study on non-orthogonal multiple access (NOMA) for NR
8. A. Benjebbour, A. Li, K. Saito, Y. Saito, Y. Kishiyama, T. Nakamura, NOMA: from concept to standardization, in *Proceedings of 2015 IEEE Conference on Standards for Communications and Networking (CSCN)* (2015), pp. 18–23
9. W. Shin, M. Vaezi, B. Lee, D.J. Love, J. Lee, H.V. Poor, Non-orthogonal multiple access in multi-cell networks: theory, performance, and practical challenges. IEEE Commun. Mag. **55**(10), 176–183 (2017)
10. M. Shirvanimoghaddam, M. Dohler, S.J. Johnson, Massive non-orthogonal multiple access for cellular IoT: potentials and limitations. IEEE Commun. Mag. **55**(9), 55–61 (2017)
11. Z. Ding, Y. Liu, J. Choi, Q. Sun, M. Elkashlan, I. Chih-Lin, H.V. Poor, Application of non-orthogonal multiple access in LTE and 5G networks. IEEE Commun. Mag. **55**(2), 185–191 (2017)
12. S.M.R. Islam, N. Avazov, O.A. Dobre, K.s. Kwak, Power-domain non-orthogonal multiple access (NOMA) in 5G systems: potentials and challenges. IEEE Commun. Surv. Tutor. **19**(2), 721–742 (2017)
13. L. Dai, B. Wang, Y. Yuan, S. Han, I. Chih-Lin, Z. Wang, Non-orthogonal multiple access for 5G: solutions, challenges, opportunities, and future research trends. IEEE Commun. Mag. **53**(9), 74–81 (2015)

14. Y. Tao, L. Liu, S. Liu, Z. Zhang, A survey: several technologies of non-orthogonal transmission for 5G. China Commun. **12**(10), 1–15 (2015)
15. Y. Wang, B. Ren, S. Sun, S. Kang, X. Yue, Analysis of non-orthogonal multiple access for 5G. China Commun. **13**(2), 52–66 (2016)
16. Z. Ding, X. Lei, G.K. Karagiannidis, R. Schober, J. Yuan, V.K. Bhargava, A survey on non-orthogonal multiple access for 5g networks: research challenges and future trends. IEEE J. Sel. Areas Commun. **35**(10), 2181–2195 (2017)
17. Z. Wei, J. Yuan, D.W.K. Ng, M. Elkashlan, Z. Ding, A survey of downlink non-orthogonal multiple access for 5G wireless communication networks. ZTE Commun. **14**(4), 17–25 (2016)
18. 3GPP R1-165021, WF on common features and general framework of MA schemes
19. 3GPP R1-165021, Performance of interleave division multiple access (IDMA) in combination with OFDM family waveforms
20. 3GPP TSG-RAN WG1-163992, Non-orthogonal multiple access candidate for NR
21. 3GPP R1-162385, Multiple access schemes for new radio interface
22. 3GPP R1-164329, Initial LLS results for UL non-orthogonal multiple access
23. 3GPP R1-164869, Low code rate and signature based multiple access scheme for NR
24. 3GPP R1-162226, Discussion on multiple access for new radio interface
25. 3GPP R1-162517, Considerations on DL/UL multiple access for NR
26. 3GPP R1-165019, Non-orthogonal multiple access for NR
27. 3GPP R1-163111, Initial views and evaluation results on non-orthogonal multiple access for NR uplink
28. 3GPP R1-163383, Candidate solution for new multiple access
29. 3GPP R1-167535, New uplink non-orthogonal multiple access schemes for NR
30. 3GPP R1-163510, Candidate NR multiple access schemes
31. 3GPP R1-164346, MA for eMBB in mmWave spectrum
32. 3GPP R1-162153, Overview of non-orthogonal multiple access for 5G
33. 3GPP RWS-150051, 5G vision for 2020 and beyond
34. K. Higuchi, A. Benjebbour, Non-orthogonal multiple access (NOMA) with successive interference cancellation for future radio access. IEICE Trans. Commun. **98**(3), 403–414 (2015)
35. A. Li, A. Benjebbour, X. Chen, H. Jiang, H. Kayama, Uplink non-orthogonal multiple access (NOMA) with single-carrier frequency division multiple Access (SC-FDMA) for 5G systems. **E98.B**, 1426–1435 (2015)
36. Z. Wei, D.W.K. Ng, J. Yuan, H.M. Wang, Optimal resource allocation for power-efficient MC-NOMA with imperfect channel state information. IEEE Trans. Commun. **65**(9), 3944–3961 (2017)
37. Y. Saito, A. Benjebbour, Y. Kishiyama, T. Nakamura, System-level performance evaluation of downlink non-orthogonal multiple access (NOMA), in *Proceedings of 2013 IEEE 24th Annual International Symposium on Personal, Indoor, and Mobile Radio Communications (PIMRC)* (2013), pp. 611–615
38. A. Jalali, R. Padovani, R. Pankaj, Data throughput of CDMA-HDR a high efficiency-high data rate personal communication wireless system, in *Proceedings of 2000 IEEE 51st Vehicular Technology Conference Proceedings*, vol. 3 (2000), pp. 1854–1858
39. N. Otao, Y. Kishiyama, K. Higuchi, Performance of non-orthogonal access with SIC in cellular downlink using proportional fair-based resource allocation, in *Proceedings of 2012 International Symposium on Wireless Communication Systems (ISWCS)* (2012), pp. 476–480
40. X. Chen, A. Benjebbour, A. Li, A. Harada, Multi-user proportional fair scheduling for uplink non-orthogonal multiple access (NOMA), in *Proceedings of 2014 IEEE 79th Vehicular Technology Conference (VTC Spring)* (2014), pp. 1–5
41. M. Kobayashi, G. Caire, An iterative water-filling algorithm for maximum weighted sum-rate of Gaussian MIMO-BC. IEEE J. Sel. Areas Commun. **24**(8), 1640–1646 (2006)
42. M.R. Hojeij, J. Farah, C.A. Nour, C. Douillard, New optimal and suboptimal resource allocation techniques for downlink non-orthogonal multiple access. Wirel. Person. Commun. Int. J. **87**(3), 837–867 (2016)

43. P. Parida, S.S. Das, Power allocation in OFDM based NOMA systems: a DC programming approach, in *Proceedings of 2014 IEEE Globecom Workshops (GC Wkshps)* (2014), pp. 1026–1031
44. A. Benjebbovu, A. Li, Y. Saito, Y. Kishiyama, A. Harada, T. Nakamura, System-level performance of downlink NOMA for future LTE enhancements, in *Proceedings of 2013 IEEE Globecom Workshops (GC Wkshps)* (2013), pp. 66–70
45. N. Ye, A. Wang, X. Li, W. Lin, X. Hou H, Yu, On constellation rotation of NOMA With SIC receiver. IEEE Commun. Lett. **22**(3), 314–517 (2018)
46. J. An, K. Yang, J. Wu, N. Ye, S. Guo, Z. Liao, Achieving sustainable ultra-dense heterogeneous networks for 5G. IEEE Commun. Mag. **55**(12), 84–90 (2017)
47. Y. Fu, Y. Chen, C.W. Sung, Distributed power control for the downlink of multi-cell NOMA systems. IEEE Trans. Wirel. Commun. **16**(9), 6207–6220 (2017)
48. L. Ping, L. Liu, K. Wu, W.K. Leung, Interleave division multiple-access. IEEE Trans. Wirel. Commun. **5**(4), 938–947 (2006)
49. L. Ping, Q. Guo, J. Tong, The OFDM-IDMA approach to wireless communication systems. IEEE Wirel. Commun. **14**(3), 18–24 (2007)
50. L. Ping, L. Liu, K.Y. Wu, W.K. Leung, Approaching the capacity of multiple access channels using interleaved low-rate codes. IEEE Commun. Lett. **8**(1), 4–6 (2004)
51. H. Wu, L. Ping, A. Perotti, User-specific chip-level interleaver design for IDMA systems. Electron. Lett. **42**(4), 233–234 (2006)
52. R. Zhang, L. Hanzo, Three design aspects of multicarrier interleave division multiple access. IEEE Trans. Veh. Technol. **57**(6), 3607–3617 (2008)
53. L. Ping, L. Liu, K.Y. Wu, W.K. Leung, On interleave-division multiple-access, in *Proceedings of 2004 IEEE International Conference on Communications*, vol. 5 (2004), pp. 2869–2873
54. H. Nikopour, H. Baligh, Sparse code multiple access, in *Proceedings of 2013 IEEE 24th Annual International Symposium on Personal, Indoor, and Mobile Radio Communications (PIMRC)* (2013), pp. 332–336
55. K. Au, L. Zhang, H. Nikopour, E. Yi, A. Bayesteh, U. Vilaipornsawai, J. Ma, P. Zhu, Uplink contention based SCMA for 5G radio access, in *Proceedings of 2014 IEEE Globecom Workshops (GC Wkshps)* (2014), pp. 900–905
56. H. Yu, Z. Fei, N. Yang, N. Ye, Optimal design of resource element mapping for sparse spreading non-orthogonal multiple access. IEEE Wirel. Commun. Lett. **PP**(99), 1 (2018)
57. M. Taherzadeh, H. Nikopour, A. Bayesteh, H. Baligh, SCMA codebook design, in *Proceedings of 2014 IEEE 80th Vehicular Technology Conference (VTC2014-Fall)* (2014), pp. 1–5
58. B. Ren, Y. Wang, X. Dai, K. Niu, W. Tang, Pattern matrix design of PDMA for 5G UL applications. China Commun. **13**(S2), 159–173 (2016)
59. J. Zeng, B. Li, X. Su, L. Rong, R. Xing, Pattern division multiple access (PDMA) for cellular future radio access, in *Proceedings of 2015 International Conference on Wireless Communications Signal Processing (WCSP)* (2015), pp. 1–5
60. S. Chen, B. Ren, Q. Gao, S. Kang, S. Sun, K. Niu, Pattern division multiple access—A novel nonorthogonal multiple access for fifth-generation radio networks. IEEE Trans. Veh. Technol. **66**(4), 3185–3196 (2017)
61. P. Li, Y. Jiang, S. Kang, F. Zheng, X. You, Pattern division multiple access with large-scale antenna array (2017)
62. J. Zeng, B. Liu, X. Su: Interleaver-based pattern division multiple access with iterative decoding and detection, in *Proceedings of 2017 IEEE Vehicular Technology Conference* (2017), pp. 1–5
63. Z. Yuan, G. Yu, W. Li, Y. Yuan, X. Wang, J. Xu, Multi-user shared access for internet of things, in *Proceedings of 2016 IEEE Vehicular Technology Conference* (2016), pp. 1–5
64. H. Hu, J. Wu, New constructions of codebooks nearly meeting the Welch bound with equality. IEEE Trans. Inf. Theory **60**(2), 1348–1355 (2014)
65. X. Meng, Y. Wu, Y. Chen, M. Cheng, Low complexity receiver for uplink SCMA system via expectation propagation, in *Proceedings of 2017 IEEE Wireless Communications and Networking Conference (WCNC)* (2017), pp. 1–5
66. 3GPP R1-164268, GB and GF MA for mMTC

67. 3GPP R1-166403, Grant-free multiple access schemes for mMTC
68. 3GPP R1-165021, WF on clarification of grant-free transmission for mMTC
69. 3GPP R1-1609398, Uplink grant-free access for 5G mMTC
70. 3GPP R1-167392, Discussion on multiple access for UL mMTC
71. 3GPP R1-166405, Discussion on grant-free concept for UL mMTC
72. N. Ye, A. Wang, X. Li, H. Yu, A. Li, H. Jiang, A random non-orthogonal multiple access scheme for mMTC, in *Proceedings of 2017 IEEE 85th Vehicular Technology Conference* (2017), pp. 1–6
73. Z. Sun, Y. Xie, J. Yuan, T. Yang, Coded slotted ALOHA for erasure channels: design and throughput analysis. IEEE Trans. Commun. **65**(11), 4817–4830 (2017)
74. E. Paolini, C. Stefanovic, G. Liva, P. Popovski, Coded random access: applying codes on graphs to design random access protocols. IEEE Commun. Mag. **53**(6), 144–150 (2015)
75. G. Liva, Graph-based analysis and optimization of contention resolution diversity slotted aloha. IEEE Trans. Commun. **59**(2), 477–487 (2011)
76. L. Toni, P. Frossard, Prioritized random MAC optimization via graph-based analysis. IEEE Trans. Commun. **63**(12), 5002–5013 (2015)
77. G. Liva, E. Paolini, M. Lentmaier, M. Chiani, Spatially-coupled random access on graphs, in *Proceedings of 2012 IEEE International Symposium on Information Theory Proceedings* (2012), pp. 478–482
78. S. Kudekar, T.J. Richardson, R.L. Urbanke, Threshold saturation via spatial coupling: why convolutional LDPC ensembles perform so well over the BEC. IEEE Trans. Inf. Theory **57**(2), 803–834 (2011)
79. M. Ivanov, F. Brännström, A. Graell i Amat, G. Liva, Unequal error protection in coded slotted ALOHA. IEEE Wirel. Commun. Lett. **5**(5), 536–539 (2016)
80. D. Jia, Z. Fei, H. Lin, J. Yuan, J. Kuang, Distributed decoding for coded slotted aloha. IEEE Commun. Lett. **21**(8), 1715–1718 (2017)
81. Č. Stefanović, P. Popovski, Coded slotted ALOHA with varying packet loss rate across users, in *Proceedings of 2013 IEEE Global Conference on Signal and Information Processing* (2013), pp. 787–790
82. Z. Sun, L. Yang, J. Yuan, M. Chiani, A novel detection algorithm for random multiple access based on physical-layer network coding, in *Proceedings of 2016 IEEE International Conference on Communications Workshops (ICC)* (2016), pp. 608–613
83. C. Stefanovic, P. Popovski, ALOHA random access that operates as a rateless code. IEEE Trans. Commun. **61**(11), 4653–4662 (2013)
84. B. Wang, L. Dai, Y. Yuan, Z. Wang, Compressive sensing based multi-user detection for uplink grant-free non-orthogonal multiple access, in *Proceedings of 2016 IEEE Vehicular Technology Conference* (2016), pp. 1–5
85. A. Bayesteh, E. Yi, H. Nikopour, H. Baligh, Blind detection of SCMA for uplink grant-free multiple-access. Int. Sym. Wirel. Commun. Syst. 853–857 (2014)
86. G. Chen, J. Dai, K. Niu, C. Dong, Sparsity-inspired sphere decoding (SI-SD): a novel blind detection algorithm for uplink grant-free sparse code multiple access. IEEE Access. **PP**(99), 1 (2017)
87. J. Liu, G. Wu, S. Li, O. Tirkkonen, Blind detection of uplink grant-free SCMA with unknown user sparsity, in *Proceedings of 2017 IEEE International Conference on Communications* (2017), pp. 1–6
88. 3GPP R1-1612573, Collision analysis of grant-free based multiple access
89. 3GPP R1-1608919, Considerations on pre-configured resource for grant-free based UL non-orthogonal MA
90. X. Chen, Z. Zhang, C. Zhong, R. Jia, D.W.K. Ng, Fully non-orthogonal communication for massive access. IEEE Trans. Commun. **PP**(99), 1 (2017)
91. 3GPP R1-1609227, On MA resource and MA signature configurations
92. 3GPP R1-1608917, Considerations on random resource selection
93. 3GPP R1-1609647, On MA resources for grant-free transmission
94. 3GPP R1-1608859, The retransmission and HARQ schemes for grant-free

95. 3GPP R1-1609039, HARQ operation for grant-free based multiple access
96. 3GPP R1-1609649, Grant-free retransmission with diversity and combining for NR
97. 3GPP R1-1609648, Collision handling for grant-free
98. 3GPP R1-1609654, Link adaptation for grant-free transmissions
99. 3GPP R1-1610374, Support of link adaptation for UL grant-free NOMA schemes
100. S. Sesia, I. Touñk, M. Daker, LTE, *The UMTS Long Term Evolution: from Theory to Practice* (Wiley Publishing, 2009)

Chapter 3
Multiple Access Towards Non-terrestrial Networks

Abstract In this chapter, we discuss the multiple access towards non-terrestrial networks. Section 3.1 introduces the current research directions of IoT and the motivation of proposing non-territorial IoT. Section 3.2 presents the application scenarios of IoT and introduces two non-terrestrial IoT technical proposals, namely satellite IoT and unmanned aerial vehicle (UAV) IoT. Sections 3.3 and 3.4 present the physical layer technologies and non-physical Layer technologies of Satellite IoT, respectively. Section 3.5 discusses the communication technologies of UAV IoT. Section 3.6 gives a conclusion of this chapter, followed by the current challenges.

3.1 Introduction

With the digital transformation of traditional economy, Internet of things (IoT), as a new generation of communication infrastructure, is the key enabling factor of ubiquitous connectivity. It realizes ubiquitous social services, builds a beautiful vision of a Smart City, and creates a brand new network environment for human development and civilization evolution through ubiquitous access and information sharing. By 2025, IoT applications are projected to grow dramatically, reaching 27 billion connections worldwide. Therefore, as an IoT application scenario, Smart City is bound to become the direction of future urban development. The deployment of Smart City in large cities will eliminate the information islands and information chimneys existing in traditional urban development and realize the great collaboration and development of the whole industry and society.

Wireless communication technology is the core technology of enabling IoT. As a summary of existing technologies, Table 3.1 summarizes the current research directions of IoT, which mainly focuses on the standards, security, scenarios, protocols, data computing and other aspects of terrestrial IoT, and the research has been very extensive. However, with the emergence of some new type of business for the IoT, such as hydrological monitoring and environmental protection, terrestrial wireless communication has been unable to meet requirements. Therefore, new ideas are

Table 3.1 Summary of existing surveys about IoT

Classification	Direction
Territorial IoT	Standard [1]
	Overview [2, 3]
	Security [4–6]
	Data collection [7–9]
	Scenarios [10–12]
	Protocols [13, 14]
	Data calculation [15–17]
	Integrating with 5G [18]
Non-territorial IoT	UAV [19]
	Satellite [20]

urgently needed to supplement the deficiency of terrestrial wireless communication. The wide coverage and low cost of non-terrestrial wireless communications complement terrestrial wireless communications.

3.2   Overview on Non-terrestrial IoT

IoT is primarily for device-to-device (D2D) communication. Different from cellular communication, IoT requires a lower transmission rate, a higher number of device connections, and requires low-power transmission due to limited power of the terminal device. In general, IoT application scenarios can be divided into two parts in Smart City: 1. Small range and short distance transmission scenarios; 2. Large range and long distance transmission scenarios. For the terminals in the first scenario, they are usually covered by local area network (LAN), such as WiFi, ZigBee, etc. The terminals in the latter scenario are covered by low power wide area network (LPWAN). Typical applications of these scenarios are shown in Table 3.2.

IoT technologies for different scenarios have different characteristics. Figure 3.1 illustrates the relationship between coverage and throughput for different technologies. Figure 3.2 shows the relationship between coverage and equipment life for LAN

Table 3.2 Typical applications of scenarios

Scenarios	Typical application	Appropriate technology
Small range and short distance transmission	Smart home, E-health	LAN
Wide range and long distance transmission	Smart city, smart grid, disaster warning	LPWAN

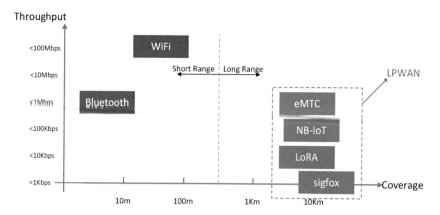

Fig. 3.1 Relationship between coverage and throughput

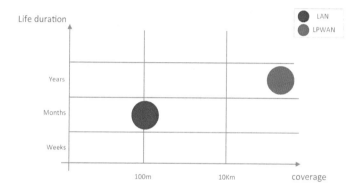

Fig. 3.2 Relationship between coverage and life duration

and LPWAN. As can be seen from these two figures, LPWAN has low throughput but long terminal battery life, and LAN requires higher throughput and does not requires strict power dissipation limits. Since this chapter focuses on Smart City, LPWAN is the key point.

Table 3.3 shows the relevant typical technologies and important indicators contained in LPWAN.

The application scenario of LPWAN can be divided into four parts, as shown in Table 3.4. To cover a wide range of scenario with a small number of terminals in some big mountain cities, considering the relatively limited coverage of the terrestrial base station, the complex geographical conditions, and the limited and dispersed equipment, the base station is not only difficult to implement, but also can only access fewer IoT equipment, which makes the construction and communication costs surge. Significant contrast with the terrestrial wireless communications, the non-terrestrial wireless communications has unique advantages, as follows:

Table 3.3 Technical indicators of LPWAN

Feature	LoRa	NB-IoT (R13+)	LTE-M (R13)
Modulation	Chirp SS	OFDMA	OFDMA
Rx bandwidth	500–125 KHz	200 KHz	20–1.4 MHz
Max output power	20 dBm	20 dBm	23/30 dBm
Data rate	290 bps–50 Kbps	~20 Kbps	200 Kbps–1 Mbps
Link budget	154 dB	150 dB	146 dNB
Power efficiency	No. 1	No. 2	No. 3

Table 3.4 Subdivided scenes of LPWAN

Scenarios	Application
Small amount of terminals and large-scale distribution	Tsunami warning, landslide warning
Small amount of terminals and small range distribution	Military application
Large number of terminals and large-scale distribution	Warehousing, dock
Large amount of terminals and small range distribution	Gas, water

- Wide coverage, which can realize global coverage, and sensor laying is not limited by space;
- Not affected by extreme terrain;
- Strong destruction resistance of the system.

These advantages totally fit into the scenario of large coverage and few terminals. In General, non-terrestrial IoT communication platforms can be considered to include satellite platforms and UAV platforms. However, the characteristics of satellite IoT and UAV IoT are different, and so are the technical requirements. Therefore, we will study these two technical approaches separately in the following paper.

3.2.1   Satellite IoT

Since the features of satellite orbit make different kinds of satellites have different characteristics, it is very important to choose the right kind of satellites to meet the needs of IoT. The geostationary earth orbit (GEO) satellite is stationary to the Earth and can provide continuous service to an area. However, it is in a higher orbit resulting in large round-trip delay and signal attenuation. In contrast to GEO, The low Earth orbit (LEO) satellite moves relative to the Earth, has a lower orbit, lower signal delay, and relatively small signal attenuation.

Due to the low cost and small size of terrestrial terminals facing Smart City, compared with GEO satellites, LEO satellites have less signal attenuation and are easier to meet the characteristics of terrestrial terminals. Therefore, there is more extensive research on the LEO satellite IoT. On the other hand, GEO satellites have the advantage of ensuring continuous service within designated coverage and providing global access beyond the polar regions with just three satellites.

However, to meet IoT requirements, satellite IoT still faces many problems:

- As IoT requires a large number of terminal access systems, wireless access technologies that avoid/tolerate collisions are required;
- Low power consumption and high energy efficiency design;
- Designed for coping with large dynamic channels;
- Deal with the scarcity of spectrum;
- Design the antenna required by satellite IoT;
- Satellite IoT protocols, standards, etc.

Oriented by requirements, the following sections describe technologies of satellite IoT in two parts: physical and non-physical, with an emphasis on technologies of physical layer.

3.2.2   UAV IoT

UAVs are highly maneuverable. Therefore, the flexible layout and mobility of UAVs are the most important features of UAV Internet of things. However, there are still many problems to overcome:

- Flexible UAV deployment to achieve efficient coverage and data collection;
- Low power consumption design;
- Coping with data collision;
- Design for dynamic channel characteristics such as Doppler frequency shift;
- Others: such as throughput optimization, low delay design, etc.

In Sect. 3.5, we focus on the flexible deployment of UAVs and summarize the technical status quo of UAV IoT oriented by requirements.

3.3   Physical Layer Technologies of Satellite IoT

A typical satellite IoT system model is shown in Fig. 3.3. The terrestrial IoT device first uploads data to a satellite relay, which then sends the data to a terrestrial data processing center for unified processing. Based on this typical system model, this

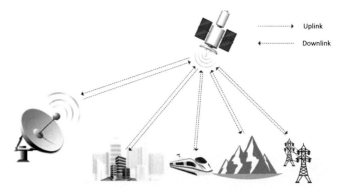

Fig. 3.3 Typical system model of satellite IoT

chapter summarizes the current research achievements on physical layer of the satellite IoT in Table 3.5 according to the technical requirements mentioned in the previous section. These technologies of physical layer mainly focus on modulation, multiple access, channel coding, frame structure design, resource scheduling, spectrum, etc., which are subsequently described in Sects. 3.3 and 3.4 later.

3.3.1 Wireless Access Technologies

To deal with the large number of terrestrial terminals access to satellites, related scholars put forward a variety of access technologies to avoid or tolerate collisions. For IoT based on GEO satellite, Hofmann et al. [21] proposed a new chirp-spread spectrum (CSS) based modulation and signaling scheme. Unipolar codes are used in the proposed transport structures in a new way, which allows a large number of devices to access a common channel randomly.

Non-orthogonal multiple access(NOMA) is an important means of terrestrial wireless access and a hotspot of B5G/6G research and standardization. Furthermore, in IoT access for integrated air-ground network, it is still a key research direction [22]. Ding et al. [23] proposed a MIMO-NOMA scheme for IoT, which uses precoding to make one user strictly meet the service quality, and the other user gets the services through NOMA opportunistically. Hu et al. [24] proposed constellation coding for multiuser reuse, which improves spectral efficiency and the number of users' access by transmitting multiple users signals simultaneously on the same frequency band. Specifically, the authors first give each user a specific number and corresponding constellation, and then they map each combination of user constellation points to a higher-order constellation point, and the schematic diagram is shown in Fig. 3.4. At the receiver, after the user detects the high-order modulation signal, each user applies the corresponding constellation decoding to get its own signal.

Table 3.5 Summary of physical layer technology of satellite IoT

Requirements	References	Physical layer technology
Wireless access technology	Hofmann and Knopp [21]	Modulation
	Yang et al. [22], Ding et al. [23], Hu et al. [24], Abramson [25], Roberts [26], Choudhury and Rappaport [27], Casini et al. [28], Ghanbarinejad and Schlegel [29]	Access
	De Gaudenzi et al. [30], Makrakis and Murthy [31], Paolini et al. [32], Yu et al. [33], Wang et al. [34], Bai and Ren [35], Wang et al. [36], Bai and Ren [37]	
	Zhao et al. ch3zhao2019random, Herrero and De Gaudenzi [39], Zhao et al. ch3zhao2018multisatellite, Zhen et al. ch3zhen2019optimal, Anteur et al. [42], Chelle et al. [43], Gan et al. [44], Ye et al. [45]	
	Kawamoto et al. [46]	Access control
	Cluzel et al. [47]	Performance analysis
High efficiency	Doré and Berg [48]	Frame structure
	Sun et al. [49]	Resource allocation
Large dynamic channel	Doroshkin et al. [50], Wu et al. [51]	Modulation
	Qian et al. [52], Qian et al. [53]	
	Perry et al. [55, 56]	Channel coding
	Cluzel et al. [57], Colavolpe et al. [58]	Receiver design
	Kodheli et al. [59]	Resource allocation
	Kodheli et al. [60]	Scheduling
Scarce spectrum	Wang et al. [61], Liang et al. [62]	Mm Wave
	Xu et al. [63]	Performance analysis
Other enabling technologies	Sanil et al. [64], Takahashi et al. [65], Jin et al. [66], Song et al. [67], Roy et al. [68]	

Random access technology is a kind of multiple access technology based on competition. Multiple users can occupy channel resources in a competitive way without signaling scheduling, which is an access method suitable for the IoT. ALOHA is the first random access technology proposed by Abramson [25], which has the problem of low channel utilization. Later, Roberts [26] introduced the concept of "time slot" based on ALOHA and proposed the Slotted-ALOHA, which reduced the collision probability of packets. On the basis of Slotted-ALOHA, researchers in recent years have studied the repeated transmission of packets to find the optimal balance between packet collision and reliability. They introduced the idea of time/frequency diversity and interference cancellation, and proposed diversity slotted ALOHA (DSA) [27],

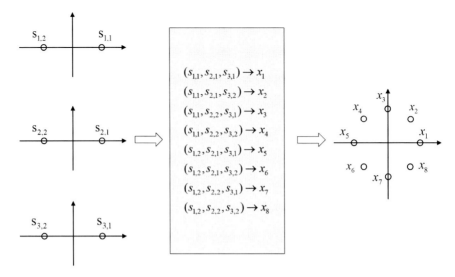

Fig. 3.4 Example of constellation mapping [24]

resolution diversity slotted ALOHA (CRDSA) [28], etc. Irregular repetition slotted ALOHA (IRSA) [29] improved CDRSA to further increase throughput. In addition, asynchronous contention resolution diversity slotted ALOHA (ACRDA) [30], spread slotted ALOHA (SSA) [31] and coded slotted ALOHA (CSA) [32] is also an improved scheme based on ALOHA. ACRDA overcomes the disadvantage of multi-terminal synchronization based on CRDSA. SSA introduces spread spectrum technology in Slotted-ALOHA; CSA introduces the combination of packet erasure correcting codes and successive interference cancellation (SIC) into Slotted-ALOHA. Irregular repetition spatially-coupled slotted ALOHA(IRSC-SA) [33] introduces the concept of spatial coupling in slotted ALOHA and applies a new density evolution (DE) method to address the unequal protection of different users. Subsequently, R-CSA [34] further improved CSA to cope with the Aloha collision problem in the multi-receiver satellite IoT. Bai et al. [35] proposed a new adaptive packet-length assisted slotted aloha scheme to cope with the large dynamic satellite channel environment. Ren's team introduced the ideas of non-orthogonal multi-access and polarized transport into slotted ALOHA, and proposed two random access methods, namely non-orthogonal slotted ALOHA (NOSA) [36] and polarized multiple input multiple output slotted Aloha (PMSA) [37], to achieve higher throughput. Zhao et al. [38] proposed a random access method called random mode multiplexing, which implements multi-user random mode access by mapping packets to a resource block (RB) composed of multiple resource elements (REs). Herrero et al. [39] proposed a multi-access scheme to provide M2M communication services for a large number of low-cost satellite-borne terminals, which improved the spectrum efficiency in the case of power imbalance.

Fig. 3.5 The process of uplink uncoordinated code domain NOMA protocol [44]

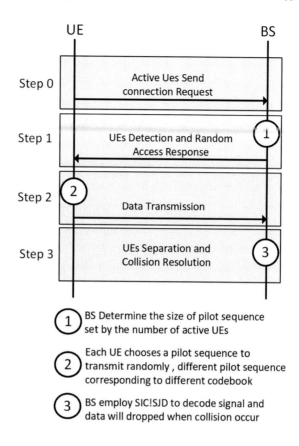

UE   BS

Step 0 — Active Ues Send connection Request

Step 1 — UEs Detection and Random Access Response ①

Step 2 ② — Data Transmission

Step 3 — UEs Separation and Collision Resolution ③

① BS Determine the size of pilot sequence set by the number of active UEs

② Each UE chooses a pilot sequence to transmit randomly, different pilot sequence corresponding to different codebook

③ BS employ SIC!SJD to decode signal and data will dropped when collision occur

Zhao et al. [40] proposed a cooperative random access scheme for multiple satellites. By using a packet structure based on single-carrier interleave frequency division multiple access (SC-IFDMA), the influence of user transmission delay on received signals of satellite nodes was overcome and the synchronization of received signals was ensured. Zhen et al. [41] proposed an enhanced spatial group-based RA scheme from the perspective of preamble design to accommodate massive and concurrent M2M random access (RA) requests and ensure human to human communication, which significantly reduces the collision probability. Time/frequency Aloha, a random access scheme suitable for ultra narrow band (UNB), is introduced by Anteur [42]. Due to the large number of devices in IoT, an overloaded random access channel (RACH) may cause a service outage. Therefore, based on the background of satellite IoT system, Chelle et al. [43] proposed a dynamic calculation method of load control parameters based on ACB. In Gan et al's article [44], the performance of an uncoordinated code domain NOMA protocol is shown in Fig. 3.5 is discussed to solve the pilot collision of massive machine type communications (mMTC) in space information network (SIN). In addition, successive interference cancellation (SIC) and successive joint decoding (SJD) were used to recover collision information under the satellite and ground channel model of shadow fading and path loss. Due to the characteristics of non-terrestrial IoT business, signaling interaction is reduced and grant-free transmission is required. To solve the collision problem caused by grant-free transmission, an rate-adaptive multiple access (RAMA) scheme was proposed [45].

Aiming at the problem that a large number of territorial IoT terminals upload data that may easily lead to data collision, Kawamoto et al. [46] divided IoT terminals into groups and allocated satellite bandwidth control access in a divide and conquer manner. In Cluzel et al's article [47], an abstract estimation method of BER and PER using physical layer is studied under a time-frequency random scheme.

3.3.2 High-Efficacy Resource Allocation

Based on the modified Zadoff-Chu sequence, Doré et al. [48] proposed an efficient channel synchronization frame structure, which has low level PAPR and is suitable for large levels Doppler.

It is worth noting that the power resources and storage resources of the satellite are limited, inefficient resource allocation may lead to interruption events, and the limited storage resources may overflow. For the communication between satellite and ground station, a low-cost source coding scheme is needed to compress the image information efficiently. Distributed source coding of hyperspectral images based on low complexity discrete cosine transform (DCT) can effectively reduce the complexity of the coding side and is suitable for on-star signal processing. In Sun et al's article [49], the authors first established a long-term joint power distribution and rate control scheme for satellite IoT NOMA downlink system. The optimal SIC decoding sequence is difficult to express because queue states and channel states constantly change from one slot to another. Therefore, deep learning-based long-term power allocation (DL-PA) is adopted to approximate SIC decoding order by training a large amount of data. The framework of the DL-PA scheme is shown in Fig. 3.6, where $Q_i(t)$ represents the queue backlog status of UE_i at slot t, $g_i(t)$ represents the channel gain of ith downlink between S and UE_i at time slot t, $s_i(t)$ represents the index of UE with the ith largest power level at time slot t, and $p_i(t)$ represents the transmit power allocated to UE_i at time slot t.

Fig. 3.6 The framework of DL-PA scheme [49]

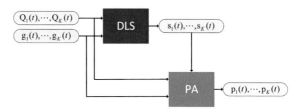

3.3.3 Large Dynamic Channel

Regarding the Doppler effect, since LoRa has not established relevant standards for the Doppler effect of fast moving satellite communications, Doroshkin et al. [50, 51] discussed the feasibility of LoRa modulation in CubeSat radio communication systems. Since chirp signal has good anti-Doppler shift performance, Qian et al. [52] studied the application of LoRA technology in LEO satellite, placing emphasis on chirp spread spectrum (CSS) modulation and introduced symmetry chirp spread spectrum (SCSS) into LEO satellite IoT. Asymmetric chirp signal (ACS) is therefore more applicable to LEO satellite IoT. Similarly, Qian et al. [53] proposed ACS. Compared with symmetric chip signal (SCS), ACS has better auto-correlation and better cross-relation in time domain and frequency domain. Yang et al. [54] proposed a folded chirp-rate shift keying (FCrSK) modulation technique, which has a strong ability to resist Doppler shift. Compared with traditional chirp-rate shift keying, its bandwidth and symbol length are consistent among chirp-rate.

Since the traditional fixed rate coding is difficult to satisfy the communication service of high-speed moving objects, the rateless coding can solve this problem to some extent. Spinal codes are a kind of rateless codes, which first converts message bits into pseudo-random sequences using a hash function and maps them to dense constellation points for transmission [55, 56]. The coding process of spinal codes is shown in Fig. 3.7, where $M_i(t)$ represents a message block, h is a random hash function, s_i represents a v-bit state.

Based on the NB-IoT standard of 3GPP, Sylvain Cluzel et al. [57] designed a kind of IoT for LEO satellite, and proposed a set of detection algorithms for Doppler shift. Similarly, Giulio Colavope et al. [58] mainly studied LoRa waveform characteristics and signal detection for LEO satellite IoT application. Then, the authors proposed a new receiver structure that exploits interference cancellation to deal with the Doppler shift.

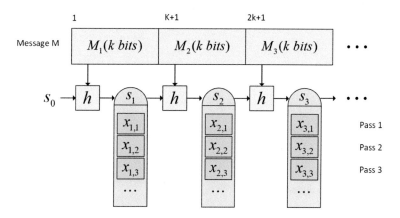

Fig. 3.7 Encoding process of spinal codes [55, 56]

NB-IoT was used to enable LEO satellite IoT [59]. In this system, the authors found the problem of high differential Doppler between different user channels. Aiming at this problem, the authors put forward a resource allocation scheme. The authors [60] proposed an uplink scheduling technique to keep differential Doppler within NB-IoT acceptable limits.

3.3.4 MmWave Transmission System

At present, service demands on throughput and service quality are rising, and the traditional frequency band is difficult to meet the practical demand. Therefore, Wang et al. [61] introduced the millimeter-wave (mm-Wave) frequency band. Combined with the massive MIMO assisted by beamforming, the authors proposed an effective adaptive random-selected multi-beamforming (ARM) estimation scheme. For hybrid satellite-territorial relay network (HSTRN), Xiao et al. [62] analyzed the system performance of mm-Wave, and further studied the influence of rainfall on mm-Wave communication under non-line-of-sight conditions. In view of the common channel interference between systems caused by the spectrum sharing of terrestrial IoT and non-terrestrial IoT, Xu et al. [63] analyzed various situations of common channel interference.

3.3.5 Other Enabling Technologies

Sanil et al. [64] studied and discussed the use of satellite facilities in c-band and x-band in IoT, and designed a multi-band microstrip antenna with three different frequencies in C-band and X-band. In order to respond flexibly to traffic demand, Takahashi et al. [65] used beamforming technology to allocate power resources, and introduced a new power resource allocation method based on transmit power and multi-beam directivity fusion control. Jin et al. [66] analyzed the special application scenarios and traffic distribution characteristics of LEO satellite IoT, and then proposed a traffic simulation method based on LEO satellite IoT. To solve the satellite downlink replanning problem, Song et al. [67] proposed a combination method based on improved genetic algorithm, and took advantage of BP neural network to optimize the initial population of genetic algorithm. In addition, Roy et al. [68] proposed a symmetric chirp signal for LEO satellites, namely symmetric chirp (SC-MCR) with multi-chirp rate which improved the cross-correlation level of SC waveform and achieves better anti-interference performance than symmetric chirp (SC) signals. In addition, time domain multiplexing (SC-MCR) waveform is designed to improve the transmission rate.

3.4 Non-physical Layer Technologies of Satellite IoT

This section summarizes the contribution of network protocol, MAC layer protocols, network architecture and satellite constellation model to technical requirements of satellite IoT.

3.4.1 High-Efficacy Protocol

As the current IoT protocol is not suitable for LEO satellite IoT, Wang et al. [69] proposed a low-complexity satellite IoT protocol, which simplifies the signaling interaction process and random access process, and achieves efficient LEO satellite IoT communication. Huang et al. [70] considered the problem of collecting data from IoT gateways via LEO satellites using an energy-saving method under time-varying uplink. In LoRa based satellite IoT, Qian et al. [71] made a critical study on how to capture SCS, and proposed a new SCS acquisition method that can balance complexity and performance.

3.4.2 Ubiquitous Network Architecture

From the perspective of network architecture, considering the future development direction of satellite network, IoT network and mobile network, Wei-Che Chien et al. [72] proposed a potential heterogeneous space and terrestrial integrated network (H-STIN) architecture for purpose of integrating various system architectures and different wireless communication protocols. To achieve a satellite-ground integrated network, network function virtualization (NFV) and software defined network (SDN) can be used to improve the capacity of wireless communication systems [73]. To cope with the problem of unbalanced traffic demand and frequent link congestion of satellite Internet of things, Liu et al. [74] proposed a routing scheme for low-earth orbit satellite network, aiming to maintain global and local load balance and optimize the data transmission of IoT. To reduce the connection delay of nanosatellite constellation, Marcano et al. [75] studied random linear network coding (RLNC) and proposed an RLNC transmission scheme. Chelle et al. [76] analyzed M2M communication extensively from the perspective of satellite, and defined a new traffic modeling for M2M communication from the perspective of the satellite.

3.4.3   Other Enabling Technologies

The European Space Agency considered the constrained limited application protocol (CoAP) as the IoT application protocol suitable for collecting IoT data through satellites in machine-type communication satellite networks. Therefore, Ridha et al. [77] studied the effective configuration of CoAP, and put forward the relevant optimal design according to the characteristics of satellite link. Similarly, Manlio et al. [78] compared two common protocol stacks, CoAP and MQTT, based on the DVB-RCS2 standard. Based on the satellite VHF data exchange system (S-VDES) MAC protocol, the detection probability of ships was analyzed and derived from the Satellite VDES by Wong et al. [79]. Furthermore, This team [80] also proposed an asynchronous multichannel pure collective ALOHA MAC protocol based on a decollision algorithm (MC-CA-SA). In order to improve energy efficiency and reduce the network delay, Wang et al. [81] proposed a joint TDMA MAC protocol (SL-MAC) applicable to LEO satellite IoT. Manlio et al. [82] analyzed the main problems hindering M2M interconnection, and proposed a M2M/IoT communication protocol stack based on oneM2M standard.

3.5   Multiple Access Technologies of UAV IoT

UAVs are now widely used in many scenarios as low-altitude platforms. However, as a low-altitude communication platform, UAV is different from traditional terrestrial communication, such as high dynamic channel environment and severe non-stationarity [83]. In view of the technical requirements faced by UAV IoT, this chapter makes a list of related technologies as shown in Table 3.6.

3.5.1   Flexible Deployment and Route Planning

Due to the high maneuverability of UAV, trajectory planning is an important optimization point for UAV to achieve air-to-ground communication. For wireless sensor network scenario, in Yang et al.s' article [84], UAV first obtains data collection points from the whole sensor network, and then uses the proposed joint genetic algorithm and ant colony optimization algorithm to determine the best route between adjacent collection points. For the task of selecting an appropriate UAV for a specific IoT task, Motlagh et al. [85] designed two solutions based on the standard of optimal energy consumption, called energy-aware selection (EAS) of UAVs and delay-aware selection (DAS) of UAVs. Similarly, Mozaffari et al. [86] mainly considered saving energy for IoT devices and realizing reliable uplink communication. Therefore, a

Table 3.6 Summary of UAV IoT technical requirements

Requirements	References
Flexible deployment	Yang and Yoo [84], Motlagh et al. [85], Mozaffari et al. [86], Liu et al. [87], Lyu et al. [88], Shi et al. [89], Qi et al. [90], Huo et al. [91], Jiang et al. [92], Ye et al. [93], Abouzaid et al. [94], Liu et al. [95], Faraizadeh et al. [96]
Low power consumption design	Al-Turjman et al. [97], Gao et al. [98], Mozaffari et al. [99], Ye et al. [100], Du et al. [101], Feng et al. [102], Motlagh et al. [103], Zhan and Lai [104], Ebrahimi et al. [105]
Data collision elimination	Almasoud and Kamal [106], Ye et al. [107], Liu et al. [108], Nomikos et al. [109], Duan et al. [110]
Large dynamic channel	Ye et al. [111], Bi et al. [112], Ding and Xu [113], Zhang and Hranilovic [114], Pang et al. [115], Wang and Liu [116]
Other enabling technologies	Choi et al. [118], Zhang et al. [119], Ye et al. [120]
	Kim and Ben-Othman [121], Ji et al. [122], Abd-Elmagid and Dhillon [123], Wang et al. [124], Wang et al. [125]
	Yu et al. [127], Qin et al. [128]
	Yan et al. [129], Santos et al. [130], Handouf et al. [131], Jingcheng et al. [132], Gaur et al. [133], Yuan et al. [134]

new method for optimal mobility of UAV is proposed, which reduces the total transmission power by 56%. In terms of system construction, according to the limited energy characteristics of UAVs, Liu et al. [87] established a multi-hop D2D link in Fig. 3.8 to extend the coverage. Lyu et al. [88] attempted to realize UAVs' uplink and downlink communication in 3D space by using cellular networks. Therefore, a new 3D system model for UAVs is constructed, and a framework for analyzing UAVs' uplink/downlink 3D coverage performance is proposed. Shi et al. [89] proposed a 3D UAV trajectory design idea: multiple DC periodically fly over the IoT devices and forward the data of the IoT to base stations (BSs). Compared with static UAV deployment, the proposed scheme reduces the average road loss by 10–15 dB. Qi et al. [90] developed a 5G IoT network based on the imagination of future Smart City. The article particularly emphasized the UAV enabling the IoT in the air to achieve 3D connectivity and connected the whole world through heterogeneous intelligent devices. Technically, a layered multi-UAV architecture for team coordination and trajectory tracking is proposed. In order to integrate different UAVs into 5G network, Huo et al. [91] proposed a multi-layer and distributed UAVs hierarchical structure. Jiang et al. [92] focused on cache enabled UAVs. In order to meet the requirements of multimedia data transmission scenario on system throughput, a collaborative IoT network structure assisted by cache-enabled UAV was proposed. Ye et al. [93] studied the UAV assisted full duplex wireless IoT network under the scenario of sparse sensor distribution, in which UAV is equipped with full duplex hybrid access point (HAP). It is assumed that the signals sent by UAV can only be received by neighboring sensors. Therefore, the authors proposed a new UAV-assisted IoT network

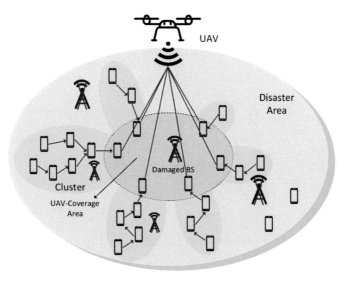

Fig. 3.8   Architecture of the UAV and D2D communication system [87]

line model to optimize system throughput. The model structure is shown in Fig. 3.9. The model shows a dynamic time division multiple access (TDMA) frame structure, which means that each sensor can only transmit information in its allocated time slot which avoids interference from other sensors. Abouzaid et al. [94] proposed an analysis model of the UAV flying mesh network, which works cooperatively in multi-hop mode to provide connectivity, data acquisition and forwarding for the terminal system, balancing end-to-end throughput and stability. Liu et al. [95] proposed a system model for UAV-enabled wireless sensor network data acquisition. In order to maximize the uplink decoding success ratio, and minimize the flight time of UAV. Farajzadeh et al. [96] adopted the power domain NOMA in the uplink, and proposed an optimization framework to determine the trade-off between numerous network parameters.

3.5.2   Low Power Consumption Design

Fadi Al-Turjman et al. [97] took UAV as the aerial base station in a specific dangerous area to enable 5G network. By optimizing the number and location of UAV, higher energy efficiency can be achieved when considering data rate, delay, throughput and other parameters. Considering the low cost and long service life of IoT devices, battery-free sensing devices based on scattering communication are a possible choice [98]. From the perspective of the system framework, A new framework was proposed by Mozaffari et al. [99] for the joint optimization of UAVs, device-UAV

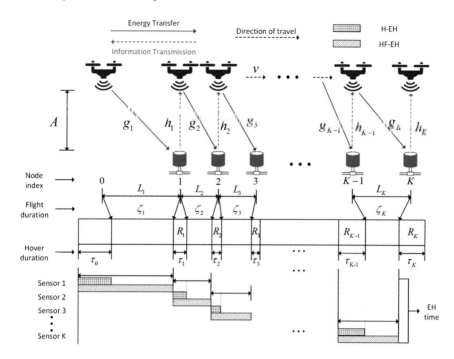

Fig. 3.9 Equivalent line model and frame structure for energy harvesting and information transmission [93]

association and uplink power control, which enables the uplink reliable transmission of IoT devices with the minimum comprehensive transmission power. In the case of multi-user overlay transmission, the mutual interference is reduced and the receiving performance is improved by designing multi-user constellation map under the simple serial interference elimination technology [100]. Du et al. [101] studied a kind of UAV as A marginal cloud to minimize UAV energy by optimizing UAV's hovering time, scheduling and resource allocation. Feng et al. [102] adopted multi-antenna UAV and multi-antenna IoT device cluster communication to form virtual MIMO link. The transmission duration and transmission power of all devices in the system are designed jointly, which greatly improves the efficiency of data acquisition of the whole flight. Considering the energy consumption and operation time of UAVs to ensure the efficient operation of value added Internet of things services (VAIoTSs), Motlagh et al. [103] proposed three complementary solutions, namely energy sensing UAV selection (EAUS), delay sensing UAV selection (DAUS) and fair weighing UAV selection (FTUS). Zhan et al. [104] studied the data collection of IoT system enabled by UAV with limited energy. In order to meet the energy budget of UAV, the authors focused on the energy of mobile propulsion system of UAV, and introduced the energy propulsion model of UAV to minimize the maximum energy consumption of all IoT devices. Aiming at the requirement of efficient data collection in dense wireless sensor networks, Ebrahimi et al. [105] proposed a projection-based compressive data gathering (CDG) method. This method aggregates data on selected projection nodes, reducing the number of data required transmissions, consequently, reducing energy consumption and extending network life.

3.5.3 Collision Resolution Design

The access of a large number of IoT devices in the region leads to a large amount of data traffic, and the spectrum is also very crowded, which makes data collisions occur from time to time. Almasoud et al. [106] studied cognitive UAV for data transmission, and data collision is effectively avoided by using spectrum sensing technology. Introducing controllable interference between multiple users to realize the increase of the number of access users is the core idea of NOMA technology [107]. Therefore, the application of NOMA technology in UAV IoT is a feasible solution to deal with multi-user collision. Liu et al. [108] used the characteristics of tolerating data collision of power domain Non-orthogonal multiple access (PD-NOMA), introducing PD-NOMA into UAV-assisted heterogeneous IoT emergency communication, so that air-to-ground (A2G) and ground-ground(G2G) communication are compatible in the same spectrum, which solved the data collision problem and further proposed a distributed SIC-free NOMA (DSF-NOMA) solution. Also using the power domain NOMA, buffering-aided (BA) relay selection was applied to the uplink of NOMA network where both user and device exist by Nomikos et al. [109], and a relay selection strategy based on dynamic decoding sequence was proposed, namely flex-NOMA. This strategy avoided the need for transmitter channel state information and reduced the probability of packet collision and delay. In addition, Duan et al. [110] combined UAV and NOMA to build a high-capacity uplink transmission system that can tolerate collisions. By jointly optimizing UAV flight height, transmission power and sub-channel allocation, the system capacity is maximized.

3.5.4 Large Dynamic Channel

Xu et al. introduced coherent/incoherent spatial modulation (SM) and space-time block coding using index shift keying (STBC-ISK) to deal with Doppler correspondence, where coherent SM and coherent STBC-ISK structure diagrams are shown in Fig. 3.10. In the complex channel environment, NOMA can be used to optimize access efficiency, reduce mutual interference and improve transmission robustness [111]. In terms of channel modeling, with the introduction of von-mises-fisher (VMF) scattering distribution, a new 3D MIMO channel model was proposed by Bi et al. [112]. Simulation results have shown that the model can accurately evaluate A2G UAV channel. Ding et al. [113] proposed a scheme called block turbo coded OFDM for high-speed UAV data link, which combines block turbo codes(BTCs)

Fig. 3.10 Schematics of SM
(**a**) and STBC-ISK (**b**) [117]

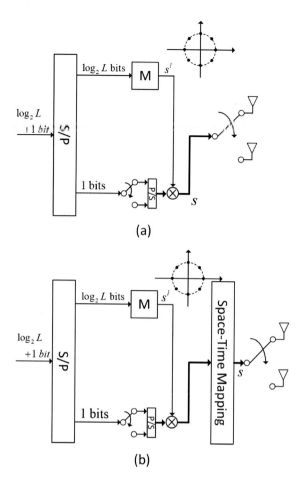

(a)

(b)

with OFDM. The traditional channel coding relies on the estimation of channel state information (CSI) and the active bit rate selection, which cannot adapt to the rapidly changing channel conditions. Rateless codes can achieve almost optimal bit rates under rapidly changing channel conditions without CSI estimation and explicit rate selection. Zhang et al. [114] and Pang et al. [115] studied the applications of rateless raptor codes and spinal codes in UAV IoT respectively. In addition, polar codes are also a promising channel encoding method for higher throughput performance in UAV IoT systems [116].

3.5.5 Other Enabling Technologies

In this subsection, the content is divided into four parts: low-latency design, joint optimization, data acquisition technology and other technologies, which will be explained in turn below.

For the UAV cluster service scenario, a UAV management system needs to be designed due to the system's demand for real-time data delivery. Choi et al. [118] designed a procedure that can realize real-time data transmission for oneM2M message flow. Aiming at the challenge posed by the high time-sensitive business of UAV IoT to effective routing, Zhang et al. [119] proposed a layered UAV swarm network structure and designed a low-latency routing algorithm (LLRA) based on partial location information and network structural connectivity. Considering the ultra-low latency requirements of the tactile IoT, grant-free NOMA is a feasible solution that can take advantage of its non-orthogonal and unlicensed transmission to achieve low-latency access [120].

Based on the reinforced barriers and collision avoidance characteristics of the heterogeneous UAV, Kim et al. [121] conducted joint optimization of the UAV's flight path without collision. Aiming at the problem that dense network signaling interactions may cause aggregation interference to terminal nodes, Ji et al. [122] proposed a joint optimization scheme that can effectively improve system throughput, interrupt probability and lawlessness by considering such parameters as transmission power, scaling factor and UAV relay selection. To ensure timely delivery of information, Abd-Elmagid et al. [123] constructed an optimization problem that takes into account UAV's flight trajectory, energy allocation and service time allocation for packet transmission, then proposed an effective iterative algorithm. Spectrum sharing in heterogeneous networks leads to cross-layer interference, which makes the power association problem in three-layer networks of satellite, UAV and macro cellular become a key problem. A two-stage joint hovering altitude and power control scheme was proposed by Wang et al. [124], which effectively improved the system throughput. In the UAV-assisted IoT communication system, Wang et al. [125] maximized the sum rate by jointly optimizing UAV altitude, user equipment uplink transmission power and user equipment-UAV scheduling.

Chakareski et al. [126] investigated the immersive data acquisition technology under the remote virtual reality scene based on UAV IoT, and designed the scalable source channel viewpoint coding, so as to maximize the reconstruction fidelity of the data captured at each UAV position at the ground convergence point. To realize accurate and efficient data sampling and reconstruction, Yu et al. [127] proposed a spatial data sampling scheme of UAV using a denoising autoencoder (DAE) neural network. Autoencoder (AE) is a neural network model for feature extraction, which has the ability to process nonlinear data. DAE evolved from basic AE, which can extract features from corrupted data and reconstruct the original data. The structure of DAE is shown in Fig. 3.11, where x represents original data, $\tilde{x} = q_D(x)$ represents the data corrupted by x, q_D is corruption function. As shown in Fig. 3.11, $y = f_\theta(\tilde{x})$ is coded data from $\tilde{x}$, $z = g_{\theta'}(y)$ represents decoded data form y. Qian et al. [128]

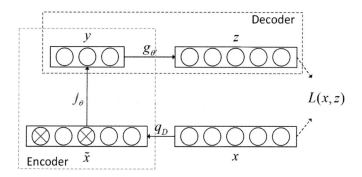

Fig. 3.11 Structure of DAE [127]

designed a novel protocol using existing ultra-low-power physical and asynchronous media access control mechanisms, as well as a lightweight application layer for data collection.

To realize the communication among various systems that may not contain the available electrical volume to support the traditional transmission of required signals, Santos et al. [130] proposed a new small electric antenna realized by direct antenna modulation (DAM). Handouf et al. [131] studied the problem of activity energy optimization of UAV, introduced the constraint optimization method of activity cycle adjustment, finally improved the system coverage and rate performance. Zhao et al. [132] proposed a UAV positioning method, which uses the range resolution ability of wide-band radar to infer the target position, and uses time-frequency analysis to analyze the Doppler effect generated by rotor rotation to determine whether it is a UAV. Gaur et al. [133] proposed an efficient vertical handover mechanism between different networks, which improves the communication reliability of beyond line of sight (BLOS) and reduces the cost. In order to achieve the optimal summation rate of UAV, Wang et al. conducted joint optimization of IoT equipment-UAV scheduling, uplink transmission power of IoT equipment and UAV altitude. Yuan et al. [134] mainly studied the super-reliable communication system of UAV swarm, designed the software protocol stack and RF hardware, and developed the open-source UAV cluster platform, namely Easy- Swarm. In the UAV-assisted IoT communication network, it is necessary to study the access selection of UAVs and the bandwidth allocation of BS to balance the network performance. Therefore, Yan et al. [129] proposed a layered game framework. The authors modeled the bandwidth allocation problem as a non-cooperative game. According to the ergodic rate analysis, this chapter proposed a game framework, i.e., a hierarchical Stackelberg game framework. The framework consists of the follower evolution game for UAVs access selection and the leader non-cooperative game for BS bandwidth allocation, as shown in Fig. 3.12.

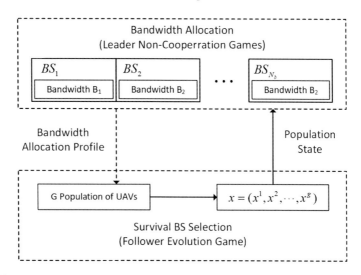

Fig. 3.12   Hierarchical Stackelberg game framework [129]

3.6   Conclusions

As a complement to the territorial IoT platforms in extreme geographic environments, non-territorial IoT platforms can effectively improve IoT coverage and enhance network reliability and availability. The chapter mainly discusses two non-territorial IoT technical paths, satellite IoT and UAV IoT. This chapter studies and sorts out the status quo of both technologies according to the perspective of technical requirements. In terms of satellite IoT, we divide existing technologies into physical and non-physical layers. As for UAV IoT, most of the papers focus on UAV layout optimization.

However, there are still some defects in the present study. As for the satellite Internet of things, it is observed that the research is mainly conducted under the LoRa system, while the OFDM system is less studied. In terms of UAV IoT, there is relatively little research on physical layer technology. In general, the research on physical layer technology for large dynamic and low power consumption is still insufficient, and the research on networking technologies for non-territorial IoT platform is relatively superficial.

References

1. A. Meddeb, Internet of things standards: who stands out from the crowd? IEEE Commun. Mag. **54**(7), 40–47 (2016)
2. A. Al-Fuqaha, M. Guizani, M. Mohammadi et al., Internet of things: a survey on enabling technologies, protocols, and applications. IEEE Commun. Surv. Tutor. **17**(4), 2347–2376 (2015)

3. J. Lin, W. Yu, N. Zhang et al., A survey on internet of things: architecture, enabling technologies, security and privacy, and applications. IEEE Internet Things J. **4**(5), 1125–1142 (2017)
4. K. Zhao, L. Ge, A survey on the internet of things security, in *Proceedings of 2013 Ninth International Conference on Computational Intelligence and Security* (IEEE, 2013), pp. 663–667
5. V. Hassija, V. Chamola, V. Saxena et al., A survey on IoT security: application areas, security threats, and solution architectures. IEEE Access **7**, 82721–82743 (2019)
6. J. Granjal, E. Monteiro, J.S. Silva, Security for the internet of things: a survey of existing protocols and open research issues. IEEE Commun. Surv. Tutor. **17**(3), 1294–1312 (2015)
7. C.W. Tsai, C.F. Lai, M.C. Chiang et al., Data mining for internet of things: a survey. IEEE Commun. Surv. Tutor. **16**(1), 77–97 (2013)
8. S. Pattar, R. Buyya, K.R. Venugopal et al., Searching for the IoT resources: fundamentals, requirements, comprehensive review, and future directions. IEEE Commun. Surv. Tutor. **20**(3), 2101–2132 (2018)
9. F. Alam, R. Mehmood, I. Katib et al., Data fusion and IoT for smart ubiquitous environments: a survey. IEEE Access **5**, 9533–9554 (2017)
10. A. Zanella, N. Bui, A. Castellani et al., Internet of Things for smart cities. IEEE Internet Things J. **1**(1), 22–32 (2014)
11. E. Ahmed, I. Yaqoob, A. Gani et al., Internet-of-things-based smart environments: state of the art, taxonomy, and open research challenges. IEEE Wirel. Commun. **23**(5), 10–16 (2016)
12. S.R. Islam, D. Kwak, M.H. Kabir et al., The Internet of Things for health care: a comprehensive survey. IEEE Access **3**, 678–708 (2015)
13. A. Rajandekar, B. Sikdar, A survey of MAC layer issues and protocols for machine-to-machine communications. IEEE Internet Things J. **2**(2), 175–186 (2015)
14. M.R. Palattella, N. Accettura, X. Vilajosana et al., Standardized protocol stack for the internet of (important) things. IEEE Commun. Surv. Tutor. **15**(3), 1389–1406 (2012)
15. W. Yu, F. Liang, X. He et al., A survey on the edge computing for the internet of things. IEEE Access **6**, 6900–6919 (2017)
16. M. Chiang, T. Zhang, Fog and IoT: an overview of research opportunities. IEEE Internet Things J. **3**(6), 854–864 (2016)
17. C. Mouradian, D. Naboulsi, S. Yangui et al., A comprehensive survey on fog computing: state-of-the-art and research challenges. IEEE Commun. Surv. Tutor. **20**(1), 416–464 (2017)
18. G.A. Akpakwu, B.J. Silva, G.P. Hancke et al., A survey on 5G networks for the internet of things: communication technologies and challenges. IEEE Access **6**, 3619–3647 (2017)
19. N.H. Motlagh, T. Taleb, O. Arouk, Low-altitude unmanned aerial vehicles-based Internet of Things services: comprehensive survey and future perspectives. IEEE Internet Things J. **3**(6), 899–922 (2016)
20. L. Chen, S. Thombre, K. Järvinen et al., Robustness, security and privacy in location-based services for future IoT: a survey. IEEE Access **5**, 8956–8977 (2017)
21. C.A. Hofmann, A. Knopp, Ultra-narrowband waveform for IoT direct random multiple access to GEO satellites. IEEE Internet Things J. 1–1 (2019)
22. K. Yang, N. Yang, N. Ye et al., Non-orthogonal multiple access: achieving sustainable future radio access. IEEE Commun. Mag. **57**(2), 116–121 (2019)
23. Z. Ding, L. Dai, H.V. Poor, Mimo-noma design for small packet transmission in the internet of things. IEEE Access **4**, 1393–1405 (2016)
24. D. Hu, L. He, J. Wu, A novel forward-link multiplexed scheme in satellite-based internet of things. IEEE Internet Things J. **5**(2), 1265–1274 (2018)
25. N. Abramson, The ALOHA system: another alternative for computer communications, in *Proceedings of the November 17–19, 1970, Fall Joint Computer Conference* (1970), pp. 281–285
26. L.G. Roberts, ALOHA packet system with and without slots and capture. SIGCOMM Comput. Commun. Rev. **5**(2), 28–42 (1975)

27. G. Choudhury, S. Rappaport, Diversity ALOHA-A random access scheme for satellite communications. IEEE Trans. Commun. **31**(3), 450–457 (1983)
28. E. Casini, R. De Gaudenzi, O.D.R. Herrero, Contention resolution diversity slotted ALOHA (CRDSA): an enhanced random access scheme for satellite access packet networks. IEEE Trans. Wirel. Commun. **6**(4), 1408–1419 (2007)
29. M. Ghanbarinejad, C. Schlegel, Irregular repetition slotted aloha with multiuser detection, in *Proceedings of 2013 10th Annual Conference on Wireless On-demand Network Systems and Services (WONS)*, pp. 201–205
30. R. De Gaudenzi, Herrero O. del Río, G. Acar et al., Asynchronous contention resolution diversity ALOHA: making CRDSA truly asynchronous. IEEE Trans. Wirel. Commun. **13**(11), 6193–6206 (2014)
31. D. Makrakis, K.M.S. Murthy, Spread slotted ALOHA techniques for mobile and personal satellite communication systems. IEEE J. Sel. Areas in Commun. **10**(6), 985–1002 (1992)
32. E. Paolini, G. Liva, M. Chiani, Coded slotted ALOHA: a graph-based method for uncoordinated multiple access. IEEE Trans. Inf. Theory **61**(12), 6815–6832 (2015)
33. H. Yu, Z. Fei, C. Cao et al., Analysis of irregular repetition spatially-coupled slotted ALOHA. Sci. China Inf. Sci. **62**(8), 80302 (2019)
34. R. Wang, P. Li, G. Cui, et al., Cooperative slotted ALOHA with reservation for multi-receiver satellite IoT networks, in *Proceedings of 2018 IEEE/CIC International Conference on Communications in China (ICCC)* (IEEE, 2018), pp. 593–597
35. J. Bai, G. Ren, Adaptive packet-length assisted random access scheme in LEO satellite network. IEEE Access **7**, 68250–68259 (2019)
36. Q. Wang, G. Ren, S. Gao et al., A framework of non-orthogonal slotted aloha (NOSA) protocol for TDMA-based random multiple access in IoT-oriented satellite networks. IEEE Access **6**, 77542–77553 (2018)
37. J. Bai, G. Ren, Polarized MIMO slotted ALOHA random access scheme in satellite network. IEEE Access **5**, 26354–26363 (2017)
38. B. Zhao, G. Ren, H. Zhang, Random pattern multiplexing for random access in IoT-oriented satellite networks. IEEE Syst. J. 1–12 (2019)
39. O.D.R. Herrero, R. De Gaudenzi, High efficiency satellite multiple access scheme for machine-to-machine communications. IEEE Trans. Aerosp. Electron. Syst. **48**(4), 2961–2989 (2012)
40. B. Zhao, G. Ren, H. Zhang, Multisatellite cooperative random access scheme in low earth orbit satellite networks. IEEE Syst. J. **13**(3), 2617–2628 (2019)
41. L. Zhen, H. Qin, Q. Zhang et al., Optimal preamble design in spatial group based random access for satellite-M2M communications. IEEE Wirel. Commun. Lett. **8**(3), 953–956 (2019)
42. M. Anteur, V. Deslandes, N. Thomas, et al., Ultra narrow band technique for low power wide area communications, in: *Proceedings of 2015 IEEE Global Communications Conference (GLOBECOM)* (IEEE, 2015), pp. 1–6
43. H. Chelle, M. Crosnier, R. Dhaou, et al., Adaptive load control for IoT based on satellite communications, in: *Proceedings of 2018 IEEE International Conference on Communications (ICC)* (IEEE, 2018), pp. 1–7
44. M. Gan, J. Jiao, L. Li, et al., Performance analysis of uplink uncoordinated code-domain NOMA for SINs, in *Proceedings of 2018 10th International Conference on Wireless Communications and Signal Processing (WCSP)* (IEEE, 2018), pp. 1–6
45. N. Ye, A. Wang, X. Li, et al., Rate-adaptive multiple access for uplink grant-free transmission. Wirel. Commun. Mob. Comput. **2018** (2018)
46. Y. Kawamoto, H. Nishiyama, Z.M. Fadlullah et al., Effective data collection via satellite-routed sensor system (SRSS) to realize global-scaled internet of things. IEEE Sens. J. **13**(10), 3645–3654 (2013)
47. S. Cluzel, M. Dervin, J. Radzik, et al., Physical layer abstraction for performance evaluation of LEO satellite systems for IOT using time-frequency ALOHA scheme, in *Proceedings of 2018 IEEE Global Conference on Signal and Information Processing (GlobalSIP)* (IEEE, 2018), pp. 1076–1080

48. J.B. Doré, V. Berg, TURBO-FSK: a 5G NB-IoT evolution for LEO satellite networks, in *Proceedings of 2018 IEEE Global Conference on Signal and Information Processing (GlobalSIP)* (IEEE, 2018), pp. 1040–1044

49. Y. Sun, Y. Wang, J. Jiao et al., Deep learning-based long-term power allocation scheme for NOMA downlink system in S-IoT. IEEE Access **7**, 86288–86296 (2019)

50. A.A. Doroshkin, A.M. Zadorozhny, O.N. Kus et al., Experimental study of LoRa modulation immunity to Doppler effect in CubeSat radio communications. IEEE Access **7**, 75721–75731 (2019)

51. T. Wu, D. Qu, G. Zhang, Research on LoRa adaptability in the LEO satellites internet of things, in *Proceedings of 2019 15th International Wireless Communications & Mobile Computing Conference (IWCMC)* (IEEE, 2019), pp. 131–135

52. Y. Qian, L. Ma, X. Liang, Symmetry chirp spread spectrum modulation used in LEO satellite internet of things. IEEE Commun. Lett. **22**(11), 2230–2233 (2018)

53. Y. Qian, L. Ma, X. Liang, The performance of chirp signal used in LEO satellite internet of things. IEEE Commun. Lett. **23**(8), 1319–1322 (2019)

54. C. Yang, M. Wang, L. Zheng et al., Folded chirp-rate shift keying modulation for LEO satellite IoT. IEEE Access **7**, 99451–99461 (2019)

55. J. Perry, P.A. Iannucci, K.E. Fleming, et al., Spinal codes, in *Proceedings of the ACM SIGCOMM 2012 Conference on Applications, Technologies, Architectures, and Protocols for Computer Communication* (ACM, 2012), pp. 49–60

56. J. Perry, H. Balakrishnan, D. Shah, Rateless spinal codes, in *Proceedings of the 10th ACM Workshop on Hot Topics in Networks* (ACM), p. 6

57. S. Cluzel, L. Franck, J. Radzik, et al., 3GPP NB-IoT coverage extension using LEO satellites, in *Proceedings of 2018 IEEE 87th Vehicular Technology Conference (VTC Spring)* (IEEE, 2018), pp. 1–5

58. G. Colavolpe, T. Foggi, M. Ricciulli, et al., Reception of LoRa signals from LEO satellites. IEEE Trans. Aerosp. Electron. Syst. 1–1 (2019)

59. O. Kodheli, S. Andrenacci, N. Maturo, et al., Resource allocation approach for differential Doppler reduction in NB-IoT over LEO satellite, in *Proceedings of 2018 9th Advanced Satellite Multimedia Systems Conference and the 15th Signal Processing for Space Communications Workshop (ASMS/SPSC)* (IEEE, 2018), pp. 1–8

60. O. Kodheli, S. Andrenacci, N. Maturo et al., An uplink UE group-based scheduling technique for 5G mMTC systems over LEO satellite. IEEE Access **7**, 67413–67427 (2019)

61. Y. Wang, Q. Li, J. Jiao et al., ARM: adaptive random-selected multi-beamforming estimation scheme for satellite-based internet of things. IEEE Access **7**, 63264–63276 (2019)

62. X. Liang, J. Jiao, B. Feng, et al., Performance analysis of millimeter-wave hybrid satellite-terrestrial relay networks over rain fading channel, in *Proceedings of 2018 IEEE 88th Vehicular Technology Conference (VTC-Fall)* (IEEE, 2018), pp. 1–5

63. D. Xu, G. Zhang, X. Ding, Analysis of co-channel interference in low-orbit satellite internet of things, in *Proceedings of 2019 15th International Wireless Communications & Mobile Computing Conference (IWCMC)* (IEEE, 2019), pp. 136–139

64. N. Sanil, P.A.N. Venkat, M.R. Ahmed, Design and performance analysis of multiband microstrip antennas for IoT applications via satellite communication, in *Proceedings of 2018 Second International Conference on Green Computing and Internet of Things (ICGCIoT)* (2018), pp. 60–63

65. M. Takahashi, Y. Kawamoto, N. Kato et al., Adaptive power resource allocation with multibeam directivity control in high-throughput satellite communication systems. IEEE Wirel. Commun. Lett. **8**(4), 1248–1251 (2019)

66. C. Jin, X. He, X. Ding, Traffic analysis of LEO satellite internet of things, in *Proceedings of 2019 15th International Wireless Communications & Mobile Computing Conference (IWCMC)* (IEEE, 2019), pp. 67–71

67. Y.J. Song, B.Y. Song, Z.S. Zhang et al., The satellite downlink replanning problem: a BP neural network and hybrid algorithm approach for IoT internet connection. IEEE Access **6**, 39797–39806 (2018)

68. A. Roy, H.B. Nemade, R. Bhattacharjee, Symmetry chirp modulation waveform design for LEO satellite IoT communication. IEEE Commun. Lett. **23**(10), 1836–1839 (2019)
69. Z. Wang, G. Cui, P. Li, W. Wang, Y. Zhang, Design and implementation of NS3-based simulation system of LEO satellite constellation for IoTs, in *2018 IEEE 4th International Conference on Computer and Communications (ICCC)* IEEE, 2018), pp. 806–810
70. H. Huang, S. Guo, W. Liang, K. Wang, A.Y. Zomaya, Green data-collection from GEO-distributed IoT networks through low-earth-orbit satellites. IEEE Trans. Green Commun. Netw. **3**(3), 806–816 (2019)
71. Y. Qian, L. Ma, X. Liang, The acquisition method of symmetry chirp signal used in LEO satellite Internet of Things. IEEE Commun. Lett. 1–1
72. W.-C. Chien, C.-F. Lai, M.S. Hossain, G. Muhammad, Heterogeneous space and terrestrial integrated networks for IoT: architecture and challenges. IEEE Netw. **33**(1), 15–21 (2019)
73. S. Sun, M. Kadoch, L. Gong, B. Rong, Integrating network function virtualization with SDR and SDN for 4G/5G networks. IEEE Netw. **29**(3), 54–59 (2015)
74. Z. Liu, J. Li, Y. Wang, X. Li, S. Chen, HGL: a hybrid global-local load balancing routing scheme for the internet of things through satellite networks. Int. J. Distrib. Sens. Netw. **13**(3), 1550147717692586 (2017)
75. N.J.H. Marcano, R.H. Jacobsen, On the delay advantages of a network coded transport layer in IoT nanosatellite constellations, in *Proceedings of ICC 2019-2019 IEEE International Conference on Communications (ICC)* (IEEE, 2019), pp. 1–6
76. H. Chelle, M. Crosnier, V. Deslandes, R. Dhaouz, A.-L. Beylot, Modelling discontinuous LEO satellite constellations: Impact on the machine-to-machine traffic and performance evaluation, in *Proceedings of 2016 8th Advanced Satellite Multimedia Systems Conference and the 14th Signal Processing for Space Communications Workshop (ASMS/SPSC)* (IEEE, 2016), pp. 1–7
77. R. Soua, M.R. Palattella, T. Engel, IoT application protocols optimisation for future integrated M2M-satellite networks, in *Proceedings of 2018 Global Information Infrastructure and Networking Symposium (GIIS)* (IEEE, 2018), pp. 1–5
78. M. Bacco, M. Colucci, A. Gotta, Application protocols enabling internet of remote things via random access satellite channels, in *Proceedings of 2017 IEEE International Conference on Communications (ICC)* (IEEE, 2017), pp. 1–6
79. D.T.C. Wong, Q. Chen, X. Peng, F. Chin, Detection probabilities for satellite VHF data exchange system with decollision algorithm and spot beam, in *Proceedings of 2018 IEEE 4th World Forum on Internet of Things (WF-IoT)* (IEEE, 2018), pp. 326–331
80. D.T.C. Wong, Q. Chen, X. Peng, F. Chin, Multi-channel pure collective ALOHA MAC protocol with decollision algorithm for satellite uplink, in *Proceedings of 2018 IEEE 4th World Forum on Internet of Things (WF-IoT)* (IEEE, 2018), pp. 251–256
81. C. Wang, L. Liu, H. Ma, D. Xia, SL-MAC: a joint TDMA MAC protocol for LEO satellites supported internet of things, in *Proceedings of 2018 14th International Conference on Mobile Ad-Hoc and Sensor Networks (MSN)* (IEEE, 2018), pp. 31–36
82. M. Bacco, L. Boero, P. Cassara, M. Colucci, A. Gotta, M. Marchese, F. Patrone, IoT applications and services in space information networks. IEEE Wirel. Commun. **26**(2), 31–37 (2019)
83. A.A. Khuwaja, Y. Chen, N. Zhao, M.-S. Alouini, P. Dobbins, A survey of channel modeling for UAV communications. IEEE Commun. Surv. Tutor. **20**(4), 2804–2821 (2018)
84. Q. Yang, S.-J. Yoo, Optimal UAV path planning: sensing data acquisition over IoT sensor networks using multi-objective bio-inspired algorithms. IEEE Access **6**, 13671–13684 (2018)
85. N.H. Motlagh, M. Bagaa, T. Taleb, UAV selection for a UAV-based integrative IoT platform, in *Proceedings of 2016 IEEE Global Communications Conference (GLOBECOM)* (IEEE, 2016), pp. 1–6
86. M. Mozaffari, W. Saad, M. Bennis, M. Debbah, Mobile internet of things: can UAVs provide an energy-efficient mobile architecture? in *Proceedings of 2016 IEEE Global Communications Conference (GLOBECOM)* (IEEE, 2016), pp. 1–6
87. X. Liu, Z. Li, N. Zhao, W. Meng, G. Gui, Y. Chen, F. Adachi, Transceiver design and multihop D2D for UAV IoT coverage in disasters. IEEE Internet Things J. **6**(2), 1803–1815 (2019)

88. J. Lyu, R. Zhang, Network-connected UAV: 3D system modeling and coverage performance analysis. IEEE Internet Things J. **6**(4), 7048–7060 (2019)

89. W. Shi, J. Li, N. Cheng, F. Lyu, Y. Dai, H. Zhou, X.S. Shen: 3D multi-drone-cell trajectory design for efficient IoT data collection, in *Proceedings of ICC 2019-2019 IEEE International Conference on Communications (ICC)* (IEEE, 2019), pp. 1–6

90. F. Qi, X. Zhu, G. Mang, M. Kadoch, W. Li, UAV network and IoT in the sky for future smart cities. IEEE Netw. **33**(2), 96–101 (2019)

91. Y. Huo, X. Dong, T. Lu, W. Xu, M. Yuen, Distributed and multi-layer UAV networks for next-generation wireless communication and power transfer: a feasibility study. IEEE Internet Things J. **6**(4), 7103–7115 (2019)

92. B. Jiang, J. Yang, H. Xu, H. Song, G. Zheng, Multimedia data throughput maximization in internet-of-things system based on optimization of cache-enabled UAV. IEEE Internet Things J. **6**(2), 3525–3532 (2018)

93. H.-T. Ye, X. Kang, J. Joung, Y.-C. Liang, Optimal time allocation for full-duplex wireless-powered IoT networks with unmanned aerial vehicle, in *Proceedings of ICC 2019-2019 IEEE International Conference on Communications (ICC)* (IEEE, 2019), pp. 1–6

94. L. Abouzaid, E. Sabir, A. Errami, H. Elbiaze, A queuing theoretic framework for flying mesh network assisted IoT environments, in *Proceedings of 2019 IEEE 5th World Forum on Internet of Things (WF-IoT)* (IEEE, 2019), pp. 882–887

95. S. Liu, Z. Wei, Z. Guo, X. Yuan, Z. Feng, Performance analysis of UAVs assisted data collection in wireless sensor network, in *Proceedings of 2018 IEEE 87th Vehicular Technology Conference (VTC Spring)* (IEEE, 2018), pp. 1–5

96. A. Farajzadeh, O. Ercetin, H. Yanikomeroglu, UAV data collection over NOMA backscatter networks: UAV altitude and trajectory optimization, in *Proceedings of ICC 2019—2019 IEEE International Conference on Communications (ICC)* (IEEE, 2019), pp. 1–7

97. F. Al-Turjman, J.P. Lemayian, S. Alturjman, L. Mostarda, Enhanced deployment strategy for the 5G drone-BS using artificial intelligence. IEEE Access **7**, 75999–76008 (2019)

98. X. Gao, D. Niyato, P. Wang, K. Yang, J. An, Contract design for time resource assignment and pricing in backscatter-assisted RF-powered networks. IEEE Wirel. Commun. Lett. **9**(1), 42–46 (2020)

99. M. Mozaffari, W. Saad, M. Bennis, M. Debbah, Mobile unmanned aerial vehicles (UAVs) for energy-efficient internet of things communications. IEEE Trans. Wirel. Commun. **16**(11), 7574–7589 (2017)

100. N. Ye, A. Wang, X. Li, W. Liu, X. Hou, H. Yu, On constellation rotation of NOMA with SIC receiver. IEEE Commun. Lett. **22**(3), 514–517 (2018)

101. Y. Du, K. Wang, K. Yang, G. Zhang, Energy-efficient resource allocation in UAV based MEC system for IoT devices, in *Proceedings of 2018 IEEE Global Communications Conference (GLOBECOM)*. (IEEE, 2018), pp. 1–6

102. W. Feng, J. Wang, Y. Chen, X. Wang, N. Ge, J. Lu, UAV-aided MIMO communications for 5G internet of things. IEEE Internet Things J. **6**(2), 1731–1740 (2018)

103. N.H. Motlagh, M. Bagaa, T. Taleb, Energy and delay aware task assignment mechanism for UAV-based IoT platform. IEEE Internet Things J. **6**(4), 6523–6536 (2019)

104. C. Zhan, H. Lai, Energy minimization in Internet-of-Things system based on rotary-wing UAV. IEEE Wirel. Commun. Lett. (2019)

105. D. Ebrahimi, S. Sharafeddine, P.-H. Ho, C. Assi, UAV-aided projection-based compressive data gathering in wireless sensor networks. IEEE Internet Things J. **6**(2), 1893–1905 (2018)

106. A.M. Almasoud, A.E. Kamal, Data dissemination in IoT using a cognitive UAV. IEEE Trans. Cogn. Commun. Netw. (2019)

107. N. Ye, H. Han, L. Zhao, A.-H. Wang, Uplink nonorthogonal multiple access technologies toward 5G: a survey. Wirel. Commun. Mob. Comput. (2018)

108. M. Liu, J. Yang, G. Gui, DSF-NOMA: UAV-assisted emergency communication technology in a heterogeneous Internet of Things. IEEE Internet Things J. **6**(3), 5508–5519 (2019)

109. N. Nomikos, E.T. Michailidis, P. Trakadas, D. Vouyioukas, T. Zahariadis, I. Krikidis, Flex-NOMA: exploiting buffer-aided relay selection for massive connectivity in the 5G uplink. IEEE Access **7**, 88743–88755 (2019)

110. R. Duan, J. Wang, C. Jiang, H. Yao, Y. Ren, Y. Qian, Resource allocation for multi-UAV aided IoT NOMA uplink transmission systems. IEEE Internet Things J. **6**(4), 7025–7037 (2019)

111. N. Ye, X. Li, H. Yu, L. Zhao, W. Liu, X. Hou, DeepNOMA: a unified framework for NOMA using deep multi-task learning. IEEE Trans. Wirel. Commun. **19**(4), 2208–2225 (2020)

112. Y. Bi, J. Zhang, X. Xu, W. Zhang, L. Zhang, Performance analysis of three-dimensional MIMO antenna arrays for UAV channel, in *Proceedings of 2018 IEEE/CIC International Conference on Communications in China (ICCC Workshops)* (IEEE, 2018), pp. 142–146

113. C. Ding, C. Xiu: Block turbo coded OFDM scheme and its performances for UAV high-speed data link, in *Proceedings of 2009 International Conference on Wireless Communications Signal Processing* (2009), pp. 1–4

114. W. Zhang, S. Hranilovic, Short-length raptor codes for mobile free-space optical channels, in *Proceedings of 2009 IEEE International Conference on Communications* (2009), pp. 1–5

115. X. Pang, M. Liu, Z. Li, Z. Jiao, S. Sun, Trust function based spinal codes over the mobile fading channel between UAVs, in *Proceedings of 2018 IEEE Global Communications Conference (GLOBECOM)* (2018), pp. 1–7

116. R. Wang, R. Liu, A novel puncturing scheme for polar codes. IEEE Commun. Lett. **18**(12), 2081–2084 (2014)

117. C. Xu, T. Bai, J. Zhang, R. Rajashekar, R.G. Maunder, Z. Wang, L. Hanzo, Adaptive coherent/non-coherent spatial modulation aided unmanned aircraft systems. IEEE Wirel. Commun. **26**(4), 170–177 (2019)

118. S.-C. Choi, I.-Y. Ahn, J.-H. Park, J. Kim, Towards real-time data delivery in one M2M platform for UAV management system, in *Proceedings of 2019 International Conference on Electronics, Information, and Communication (ICEIC)* (IEEE, 2019), pp. 1–3

119. Q. Zhang, M. Jiang, Z. Feng, W. Li, W. Zhang, M. Pan, IoT enabled UAV: network architecture and routing algorithm. IEEE Internet Things J. **6**(2), 3727–3742 (2019)

120. N. Ye, X. Li, H. Yu, A. Wang, W. Liu, X. Hou, Deep learning aided grant-free NOMA toward reliable low-latency access in tactile internet of things. IEEE Trans. Ind. Inform. **15**(5), 2995–3005 (2019)

121. H. Kim, J. Ben-Othman, A collision-free surveillance system using smart UAVs in multi domain IoT. IEEE Commun. Lett. **22**(12), 2587–2590 (2018)

122. B. Ji, Y. Li, B. Zhou, C. Li, K. Song, H. Wen, Performance analysis of UAV relay assisted IoT communication network enhanced with energy harvesting. IEEE Access **7**, 38738–38747 (2019)

123. M.A. Abd-Elmagid, H.S. Dhillon, Average peak age-of-information minimization in UAV-assisted IoT networks. IEEE Trans. Veh. Technol. **68**(2), 2003–2008 (2018)

124. J. Wang, C. Jiang, Z. Wei, C. Pan, H. Zhang, Y. Ren, Joint UAV hovering altitude and power control for space-air-ground IoT networks. IEEE Internet Things J. **6**(2), 1741–1753 (2018)

125. X. Wang, W. Feng, Y. Chen, N. Ge, Sum rate maximization for mobile UAV-aided Internet of Things communications system, in *Proceedings of 2018 IEEE 88th Vehicular Technology Conference (VTC-Fall)* (IEEE, 2018), pp. 1–5

126. J. Chakareski, UAV-IoT for next generation virtual reality. IEEE Trans. Image Process. **28**(12), 5977–5990 (2019)

127. T. Yu, X. Wang, A. Shami, UAV-enabled spatial data sampling in large-scale IoT systems using denoising autoencoder neural network. IEEE Internet Things J. **6**(2), 1856–1865 (2018)

128. Y. Qin, D. Boyle, E. Yeatman, A novel protocol for data links between wireless sensors and UAV based sink nodes, in *Proceedings of 2018 IEEE 4th World Forum on Internet of Things (WF-IoT)* (IEEE, 2018), pp. 371–376

129. S. Yan, M. Peng, X. Cao, A game theory approach for joint access selection and resource allocation in UAV assisted IoT communication networks. IEEE Internet Things J. **6**(2), 1663–1674 (2018)

130. J.P. Santos, F. Fereidoony, M. Hedayati, Y.E. Wang, High efficiency bandwidth electrically small antennas for compact wireless communication systems, in *Proceedings of 2019 IEEE MTT-S International Microwave Symposium (IMS)* (2019), pp. 83–86

131. S. Handouf, E. Sabir, M. Sadik, Energy-throughput tradeoffs in ubiquitous flying radio access network for IoT, in *Proceedings of 2018 IEEE 4th World Forum on Internet of Things (WF-IoT)* (IEEE, 2018), pp. 320–325
132. Z. Jingcheng, F. Xinru, Y. Zongkai, X. Fengtong, UAV detection and identification in the Internet of Things, in *Proceedings of 2019 15th International Wireless Communications & Mobile Computing Conference (IWCMC)* (IEEE, 2019), pp. 1499–1503
133. A.S. Gaur, J. Budakoti, C. H. Lung, A. Redmond, IoT-equipped UAV communications with seamless vertical handover, in *Proceedings of 2017 IEEE Conference on Dependable and Secure Computing* (IEEE, 2017), pp. 459–465
134. Z. Yuan, J. Jin, L. Sun, K.-W. Chin, G.-M. Muntean, Ultra-reliable IoT communications with UAVs: a swarm use case. IEEE Commun. Mag. **56**(12), 90–96 (2018)

Chapter 4
Constellation Design Technique for Multiple Access

Abstract In this chapter, we discuss the constellation design technique for multiple access. Section 4.1 introduces the motivation of applying constellation rotation to NOMA with SIC receiver and presents researches of constellation design technique. Section 4.2 establishes the system model and problem formulation of uplink NOMA transmission. Section 4.3 introduces the proposed capacity maximization scheme. Section 4.4 analyzes the achievable capacity and the asymptotic performance with massive antennas at the BS of the proposed method, respectively. Section 4.5 presents simulation results and gives a conclusion of this chapter.

4.1 Introduction

Non-orthogonal multiple access (NOMA), which superimposes multiple users' signals and applies successive interference cancellation (SIC) receiver, achieves the corner points of the multiple access channel (MAC) capacity pentagon with Gaussian-distributed alphabets, and is a promising technology for 5G [1]. However, with finite input constellation, the constellation constrained capacity (CCC) region is smaller than MAC capacity pentagon [2].

Authors in [2] introduce relative rotation between users to maximize the CCC region, however, efficient methods to derive the optimal rotation angle in realistic settings, e.g. with SIC receiver and multiple antennas, are not well studied. Authors in [3–5] proposed some constellation rotation methods to enhance the performance of NOMA with finite input constellations [3–5]. However, we find that the existing literatures aim to maximize the CCC by assuming the ideal maximum likelihood receiver, which leads to high receiving complexity. Further, the existing methods mainly deal with the distances in the composite constellation and ignores the noise, which is not optimal in most SNR regions. Also, dealing with distances in composite constellation cannot be directly generalized into scenarios with multiple antennas.

In this letter, we aim to enhance the performance of NOMA with SIC receiver via constellation rotation. A capacity maximization problem is formulated and trans-

© The Author(s), under exclusive license to Springer Nature Singapore Pte Ltd. 2023
N. Ye et al., *Multiple Access Technology Towards Ubiquitous Networks*,
https://doi.org/10.1007/978-981-19-4025-5_4

formed to an entropy maximization (MaxEnt) problem based on multivariate Gaussian mixture model (GMM). Then we efficiently solve the problem via variational approximation method [6]. Analysis and simulation results both reveal the advantage of proposed method over conventional NOMA and existing methods. The advantage is especially significant when multiplexed users experience small channel gain difference, which is an issue for conventional NOMA with SIC receiver [7].[1]

4.2 System Model and Problem Formulation

Consider an orthogonal frequency division multiplexing (OFDM) based uplink NOMA system with two synchronized users and one base station (BS), where users and the BS are equipped with t and r antennas, respectively. For brevity, we denote $i = 1$ for cell-interior user, and $i = 2$ for cell-edge user, where user-1 usually has better channel condition than user-2. We assume that the channels between the users and the BS are quasi-static [3] and global channel state information (CSI) is available at the BS. For each quasi-static interval, BS receives

$$y = H_1 s_1 + H_2 s_2 + n_0 = x_1 + x_2 + n_0 \tag{4.1}$$

where $s_i = [s_{i,1}, \cdots, s_{i,t}]^T$ represents the vector of modulation symbol on t antennas, n_0 is Gaussian noise with $\mathcal{CN}(\mathbf{0}, \sigma_0^2 I)$, and $H_i \in C^{r \times t}$ is the channel coefficient matrix between user-i and the BS with independent and identically distributed (i.i.d.) entries. H_i is fixed in each quasi-static interval, and we denote $x_i = H_i s_i$ as the received signal of user-i.

In each transmission interval, constellation rotation is applied on user-1 to enhance the performance of NOMA.[2] Given the rotation angle vector $\boldsymbol{\varphi} = [\varphi_1, \cdots, \varphi_t]$, the BS receives

$$y^\varphi = H_1 \left(s_1 \cdot e^{j\varphi} \right) + H_2 s_2 + n_0 = x_1^\varphi + x_2 + n_0, \tag{4.2}$$

where $e^{j\varphi} = [e^{j\varphi_1}, \cdots, e^{j\varphi_t}]$, and "$\cdot$" represents the element-wise multiplication. For illustration purpose, we denote RVs $X_1^\varphi \sim P_{X_1^\varphi}(x_1^\varphi)$, $X_2 \sim P_{X_2}(x_2)$, and $Y^\varphi \sim P_{Y^\varphi}(y^\varphi)$ as the signals of user-1, user-2, and the overall received signal, respectively. Define $\mathcal{X}_1^\varphi$ and $\mathcal{X}_2$ as the *received constellations* of user-1 and user-2 in each interval respectively, and $x_{1,j}^\varphi$ and $x_{2,k}$ as the j-th and k-th elements of $\mathcal{X}_1^\varphi$ and $\mathcal{X}_2$ respectively. Therefore, Y^φ follows

---

[1] Notation: We denote boldface upper case letters X as random vector (RV), and its lower case x as a realization of the RV. A r-dimensional RV X may take value $x = [x_1, \cdots, x_r]$ from set $\mathcal{X}$ with cardinality $|\mathcal{X}|$. $X \sim P_X(x)$ represents that X complies with distribution $P_X(x)$, where $P_X(x)$ denotes probability density function (PDF) of continuous RV according to X. $\mathcal{N}(\mu, \Gamma)$ and $\mathcal{CN}(\mu, \Gamma)$ represent the real and the circularly symmetric complex Gaussian RV, with mean μ and covariance matrix Γ, respectively.

[2] Power allocation is another important issue in NOMA. In this letter, we assume that suitable transmission powers have been allocated to users by the network side, and we emphasize on the design of constellation rotation.

$$P_{Y^\varphi}(y^\varphi) = \sum_{\substack{x^\varphi_{1,j} \in \mathcal{X}^\varphi_1 \\ x_{2,k} \in \mathcal{X}_2}} \frac{1}{|\mathcal{X}^\varphi_1||\mathcal{X}_2|} \mathcal{CN}(x^\varphi_{1,j} + x_{2,k}, \sigma_0^2 I), \tag{4.3}$$

where $\mathcal{X}^\varphi_1$ and $\mathcal{X}_2$ are supposed to have uniform distribution. Notice that $P_{Y^\varphi}(y^\varphi)$ is the weighted sum of complex Gaussian distributions, which exactly follows the definition of GMM. For simplicity, we represent $|\mathcal{X}^\varphi_1||\mathcal{X}_2|$ with $|\mathcal{X}|$, and represent the conditions of summation in (4.3), i.e., $(x^\varphi_{1,j} \in \mathcal{X}^\varphi_1, x_{2,k} \in \mathcal{X}_2)$, with $\mathcal{X}_{j,i}$ in the following.

Assume that SIC receiver is employed [1]. When the receiver fails to recover x^φ_1, x_2 will be severely disrupted. Therefore, the performance of NOMA with SIC is dominated by the cell-interior user [8], and we aim to enhance its performance by maximizing the mutual information (MI) between Y^φ and X^φ_1 as follows

$$\mathcal{P}1 : \max_\varphi I(Y^\varphi; X^\varphi_1), \text{ s.t. } \varphi_j \in [0, 2\pi), 1 \le j \le t, \tag{4.4}$$

where MI is a useful tool to enhance the detection accuracy in MAC [9, 10]. We solve $\mathcal{P}_1$ in the next section.

4.3 Constellation Rotation Method

4.3.1 Problem Transformation

To directly solve $\mathcal{P}1$, we have to calculate the joint distribution of two RVs (i.e., Y^φ and X^φ_1), which has a complicated expression. Hence, we transform $\mathcal{P}1$ into a MaxEnt problem, which only involves a single RV, according to Lemma 4.1.

Lemma 4.1 (Equivalence Property) *Maximizing $I(Y^\varphi; X^\varphi_1)$ is equivalent to maximizing $h(Y^\varphi)$, where $h(Y^\varphi)$ is the differential entropy of Y^φ.*

Proof See Appendix 1.

To avoid the complex-field integration in $h(Y^\varphi)$, we replace the r-dimensional complex RV Y^φ with the $2r$-dimensional real RV $\tilde{Y}^\varphi = \left[\text{Re}(Y^\varphi) \; \text{Im}(Y^\varphi)\right]$, where $h(\tilde{Y}^\varphi) = h(Y^\varphi)$. As a result, $\mathcal{P}1$ is equivalent to the MaxEnt problem

$$\mathcal{P}2 : \max_\varphi h(\tilde{Y}^\varphi), \text{ s.t. } \varphi_j \in [0, 2\pi), 1 \le j \le t, \tag{4.5}$$

where $\tilde{Y}^\varphi$ follows

$$P_{\tilde{Y}^\varphi}(\tilde{y}^\varphi) = \sum_{\mathcal{X}_{j,k}} \frac{1}{|\mathcal{X}|} \mathcal{N}(\mu^\varphi_{j,k}, \sigma_0^2 I_{2t}),$$

and is a $2r$-dimensional real GMM with $\mu^\varphi_{j,k} = [\text{Re}(x^\varphi_{1,j} + x_{2,k}) \; \text{Im}(x^\varphi_{1,j} + x_{2,k})]$.

4.3.2 Variational Approximation Method

To efficiently solve $\mathcal{P}2$, we employ variational approximation (VA) technique [6] to find a tight and closed-form approximation of the objective function of $\mathcal{P}2$ in the following.

First of all, we introduce the auxiliary parameters, termed variational parameters (VPs),

$$\boldsymbol{\phi} = \{\phi_{jk|lm} \mid j, l = (1, \cdots, |\mathcal{X}_1^{\varphi}|), \text{ and } k, m = (1, \cdots, |\mathcal{X}_2|)\},$$

into $h(\tilde{Y}^{\varphi})$, as shown in (4.6), with the constraint $\sum_{jk} \phi_{jk|lm} = 1$. Based on the VPs, a closed-form approximation of $h(\tilde{Y}^{\varphi})$ is derived in Lemma 4.2.

$$h\left(\tilde{Y}^{\varphi}\right) = -\sum_{\mathcal{X}_{l,m}} \left[\frac{1}{|\mathcal{X}|} \int \left(\left(\mathcal{N}\left(\boldsymbol{\mu}_{l,m}^{\varphi}, \sigma_0^2 \boldsymbol{I}_{2t}\right)\right) \log \left(\sum_{\mathcal{X}_{j,k}} \left[\phi_{jk|lm} \frac{\frac{1}{|\mathcal{X}|}\mathcal{N}\left(\boldsymbol{\mu}_{l,m}^{\varphi}, \sigma_0^2 \boldsymbol{I}_{2t}\right)}{\phi_{jk|lm}} \right] \right) \right) d\tilde{y}^{\varphi} \right] \quad (4.6)$$

$$\leq -\sum_{\mathcal{X}_{l,m}} \left[\frac{1}{|\mathcal{X}|} \int \left(\left(\mathcal{N}\left(\boldsymbol{\mu}_{l,m}^{\varphi}, \sigma_0^2 \boldsymbol{I}_{2t}\right)\right) \left(\sum_{\mathcal{X}_{j,k}} \left[\phi_{jk|lm} \log \frac{\frac{1}{|\mathcal{X}|}\mathcal{N}\left(\boldsymbol{\mu}_{l,m}^{\varphi}, \sigma_0^2 \boldsymbol{I}_{2t}\right)}{\phi_{jk|lm}} \right] \right) \right) d\tilde{y}^{\varphi} \right]$$
$$(4.7)$$

$$\triangleq L(\boldsymbol{\phi}, \boldsymbol{\varphi}),$$

Lemma 4.2 (Upper Bound) $h(\tilde{Y}^{\varphi})$ *is upper bounded by* $g(\boldsymbol{\varphi}) = \inf_{\boldsymbol{\phi}} L(\boldsymbol{\phi}, \boldsymbol{\varphi})$, *where* $L(\boldsymbol{\phi}, \boldsymbol{\varphi})$ *is defined in* (4.7).

Proof By applying Jensen's inequality on (4.6), we take the *log* term inside the *sum* term, and we readily have $L(\boldsymbol{\phi}, \boldsymbol{\varphi})$ which is the closed-form upper bound of $h(\tilde{Y}^{\varphi})$, as in (4.7).

Further, we denote the set $\mathbf{L} = \{L(\boldsymbol{\phi}, \boldsymbol{\varphi}) | \forall \boldsymbol{\phi}\}$. Obviously, $\mathbf{L}$ is lower bounded by $h(\tilde{Y}^{\varphi})$. According to *supremum and infimum principle*, there exist a $g(\boldsymbol{\varphi})$ which satisfies

$$h(\tilde{Y}^{\varphi}) \leq g(\boldsymbol{\varphi}) = \inf_{\boldsymbol{\phi}} L(\boldsymbol{\phi}, \boldsymbol{\varphi}) \leq L(\boldsymbol{\phi}', \boldsymbol{\varphi}), \forall \boldsymbol{\phi}'. \qquad \square$$

Remark 4.1 (Optimality) Since $g(\boldsymbol{\varphi})$ is the lower bound of $\mathbf{L}$, it is also the optimal approximation to $h(\tilde{Y}^{\varphi})$ in $\mathbf{L}$.

Denote $\mathcal{N}_{jk}$ and $\mathcal{N}_{lm}$ as the short notations of Gaussian distributions with mean $\boldsymbol{\mu}_{j,k}^{\varphi}$ and $\boldsymbol{\mu}_{l,m}^{\varphi}$, respectively. Now we are ready to approximate $\mathcal{P}2$ by

$$\mathcal{P}3 : \max_{\boldsymbol{\varphi}} g(\boldsymbol{\varphi}) \qquad (4.8)$$

where $g(\boldsymbol{\varphi})$ is given according to Theorem 4.1.

$$g(\boldsymbol{\varphi}) = -\sum_{\mathcal{X}_{l,m}} \sum_{\mathcal{X}_{j,k}} \left[\frac{1}{|\mathcal{X}|} \hat{\phi}_{jk|lm} \left(\log\left(|\mathcal{X}| \hat{\phi}_{jk|lm}\right) + D\left(\mathcal{N}_{lm} \| \mathcal{N}_{jk}\right) + h\left(\mathcal{N}\left(\boldsymbol{\mu}_{l,m}^{\varphi}, \sigma_0^2 \boldsymbol{I}_{2t}\right)\right) \right) \right] \quad (4.9)$$

Theorem 4.1 (VP Optimization) *For* $\forall \varphi$, $g(\varphi)$ *is found by (4.9), where* $\hat{\phi}_{jk|lm} = e^{-D(N_{lm}||N_{jk})}/\sum_{st} e^{-D(N_{lm}||N_{st})}$, *and* $D(N_{lm}||N_{jk}) = \frac{1}{2\sigma^2}\|\mu_{j,k}^{\varphi} - \mu_{l,m}^{\varphi}\|^2$ *is the Kullback-Leibler divergence between* N_{lm} *and* N_{jk}.

Proof The problem of finding the infimum of $L(\phi, \varphi)$ with respect to ϕ is convex (note that $L(\phi, \varphi)$ is in the form of $\phi - \log(\phi)$). Following the standard procedure of convex optimization, the Karush-Kuhn-Tucker (KKT) conditions are necessary and sufficient for deriving the optimal VP $\hat{\phi}_{jk|lm}$. Due to the limited space, we briefly illustrate the optimality of $\hat{\phi}_{jk|lm}$. Suppose VP $\tilde{\phi}_{jk|lm}$ is a function varying with $\tilde{y}^{\varphi}$. To make the equality in Jensen's inequality (4.7) holds, $\frac{N_{jk}}{\tilde{\phi}_{jk|lm}}$ must be a constant. Without loss of generality, we assume $\tilde{\phi}_{jk|lm} = c\frac{N_{jk}}{N_{lm}}$ with constant c. Denoting the expectation of $\tilde{\phi}_{jk|lm}$ over N_{lm} as $\hat{\phi}_{jk|lm}$, we have

$$\hat{\phi}_{jk|lm} = E_{N_{lm}}[\tilde{\phi}_{jk|lm}] \approx e^{-E_{N_{lm}}[\log(\frac{N_{jk}}{N_{lm}})]} = e^{-D(N_{lm}||N_{jk})}. \tag{4.10}$$

Normalizing and substituting (4.10) into (4.7), we get (4.9). □

Remark 4.2 The proof of Theorem 4.1 also illustrates the motivation and superiority of the proposed approximation. Observing that applying Jensen's inequality on the objective function in $\mathcal{P}2$ leads to approximation error, we hope that the equal sign, i.e., the equality in (4.7), can hold before and after exchanging the positions of *log* and *sum*. Therefore VP is used as an additional degree of freedom to reduce the approximation error, which is minimized according to Theorem 4.1.

Algorithm 4.1 VA-M Algorithm

Require: $k = 0$, $K > 0$, $\delta > 0$, and any feasible $\varphi^{(0)}$.
1: **repeat**
2:   **VA-step:** $\phi^{(k)} \leftarrow \arg\inf_{\phi} L(\phi, \varphi^{(k)})$.
3:   **M-step:** $\varphi^{(k+1)} \leftarrow \arg\sup_{\|\varphi-\varphi^{(k)}\|<\varepsilon^{(k)}} L(\phi^{(k)}, \varphi)$.
4:   $k \leftarrow k + 1$.
5: **until** $(K \leq k)$ or $(\|\varphi^{(k+1)} - \varphi^{(k)}\| \leq \delta)$
Ensure: $\varphi^{(k)}$.

Based on $\mathcal{P}3$, exhaustive search can be directly applied to get the feasible rotation angle. To further reduce the computational complexity in solving $\mathcal{P}3$, we develop an iterative variational approximation-maximization (VA-M) algorithm in the following Algorithm 4.1, inspired from the well-known expectation-maximization (EM) algorithm.

In the k-th iteration of VA-M algorithm, VA-step finds tight and closed-form approximation near $\varphi^{(k)}$, and M-step derives φ which maximizes $L(\phi^{(k)}, \varphi)$ by setting its derivative function to zero in the neighborhood of $\varphi^{(k)}$, i.e., $\|\varphi - \varphi^{(k)}\| < \varepsilon^{(k)}$, where $\varepsilon^{(k)}$ is the step size which controls the tradeoff between optimization accuracy and converging speed, which can be decided by Armijo rule. Near optimal rotation angle is derived by iteratively optimize ϕ and φ.

The complexity of VA-M Algorithm is affected by K, ε and the structure of problem. Nevertheless, the complexity is always bounded by $O(K)$, and only closed-form expressions are calculated in each iteration, which is lower than MAX-MIN [5].

We now summerize the proposed method as follows. To solve $\mathcal{P}_2$, closed-form expressions are necessary. To this end, we introduce auxiliary VPs in the objective function of $\mathcal{P}_2$ and derive a set of upper bound expressions, i.e., $\mathbf{L}$, via VA in Lemma 4.2. Furthermore, to increase the accuracy of approximation, we find the lower bound of $\mathbf{L}$, i.e., the most accurate upper bound of $h(\tilde{Y}^\varphi)$, in Theorem 4.1, and use it as the objective function of $\mathcal{P}_3$. Eventually, $\mathcal{P}_1$ is approximated by $\mathcal{P}_3$, whose optimal solution is obtained by Algorithm 4.1.

Remark 4.3 When N users are multiplexed, the received signal can also be modeled as a GMM, and the proposed method is still applicable.

4.4 Analysis and Discussions

In this section, we analyze the achievable capacity and the asymptotic performance with massive antennas at the BS of the proposed method, respectively.

4.4.1 Achievable Capacity with SIC Receiver

We compare the proposed method with the following three methods: random rotation method [2] (labeled as 'Random method'), where the rotation angle is randomly generated, the method based on Jensen's inequality [2] (labeled as 'Jensen method'), where auxiliary parameters are not introduced, and the MAX-MIN method [3, 5] (labeled as 'MAX-MIN method'), where the minimum distance in composite constellation is maximized. Figure 4.1a shows the achievable capacities at user-1 (i.e. $I(Y^\varphi; X_1^\varphi)$), with QPSK and Rayleigh block fading. We observe that the proposed method slightly outperforms other methods due to the introduction and the optimization of VPs. Note that the proposed method has lower computational complexity than Jensen method, which uses exhaustive search, due to VA-M algorithm. Besides, larger performance gain over Jensen method can be achieved by applying VA, when more complicated constellations are assumed [6].

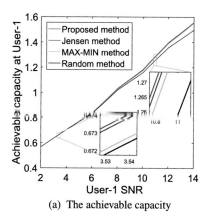

(a) The achievable capacity

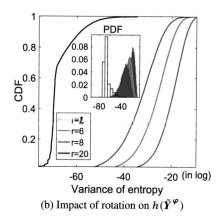

(b) Impact of rotation on $h(\tilde{Y}^{\varphi})$

Fig. 4.1 Performance analysis

4.4.2 Analysis on Infinite Number of Receiving Antenna

In uplink NOMA, multiple or even massive number of antennas may be deployed at the BS [11]. We present the following theorem to show whether the performance gain of applying rotation in NOMA still exists when $r \to \infty$.

Theorem 4.2 (Asymptotic Analysis) $\forall \varepsilon$, φ_1 and φ_2, $h(Y^{\varphi_1})$ converges to $h(Y^{\varphi_2})$ with probability, i.e.

$$\lim_{r \to \infty} P\left\{\left|h\left(Y^{\varphi_1}\right) - h\left(Y^{\varphi_2}\right)\right| < \varepsilon\right\} = 1, \tag{4.11}$$

which means that rotation does not affect $I(Y^{\varphi}; X_1^{\varphi})$.

Proof See Appendix 2.

We visually illustrate Theorem 4.2 in Fig. 4.1b. For each trial, we randomly generate **H** with $r = 2, 6, 8$, and 20. With various φ, we can calculate the variance of $h(Y^{\varphi})$. With 10^8 trials, we plot the cumulative density function of the variances. It is observed that, the entropy tends to be a constant with respect to φ with the increase of r, which reveals that rotation is not necessary when massive antennas are deployed at the BS.

4.5 Simulation Results and Conclusions

This section evaluates the uncoded bit error rate (BER) performance of the proposed method. We fix the large scale fading coefficient for each user, and assume i.i.d. Rayleigh block fading.

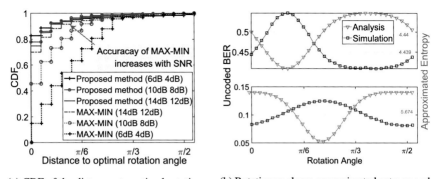

(a) CDF of the distances to optimal rotation an- (b) Rotation angle vs. approximated entropy and
gles with SISO   BER with SIMO, $r = 2$

Fig. 4.2 Performance of the proposed method

In Fig. 4.2a, we depicts the CDF of distances between optimal rotation angle, derived by solving $\mathcal{P}2$ numerically, and the angles derived by the proposed method and the MAX-MIN method [5], respectively, with QPSK input and quantization interval $\pi/32$. Results show that MAX-MIN method is inferior to the proposed method especially in low SNR. Figure 4.2b illustrates that BER is minimized when the approximated entropy $g(\varphi)$ reaches the peak. This indicates that $\mathcal{P}3$ is a good approximate problem to minimize BER.

Figures 4.3 and 4.4 show the BER of user-1 with the proposed scheme and the existing methods, with various SNR gaps between users in both SISO and SIMO scenarios, respectively. Each data block contains 168 symbols and is repeated for 10^5 times. MMSE-SIC detector is applied. It is seen that, the proposed NOMA shows significant performance gain in all SNR regions, and the gain is especially remarkable when SNR of two users are close. This implies that the proposed scheme can enhance the performance when power allocation cannot ensure significant power difference between NOMA paired users. In the left bottom of the figures, we show the convergence behavior of VA-M, with SNR equals to 8dB and SNR gap = 0 dB, which both indicate that VA-M converges fast. Finally, we note that the proposed method still works when users are not strictly synchronized.

In conclusion, the proposed constellation rotation method achieves better capacity and BER performance than conventional NOMA, especially when SNR gap among users is small.

Appendix 1

According to the definition of mutual information, we have

$$I\left(Y^{\varphi}; X_1^{\varphi}\right) = h(Y^{\varphi}) - E_{X_1^{\varphi}}\left[h\left(Y^{\varphi}|X_1^{\varphi} = x_1^{\varphi}\right)\right]$$

Fig. 4.3 Uncoded BER performance of user-1 with SISO

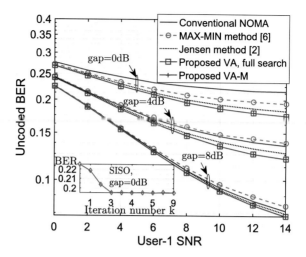

Fig. 4.4 Uncoded BER performance of user-1 with SIMO, $r = 2$

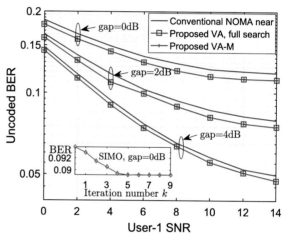

$$= h(Y^{\varphi}) - E_{X_1^{\varphi}} \left[h \left(x_1^{\varphi} + X_2 + N_0 \right) \right]$$
$$\overset{(a)}{=} h(Y^{\varphi}) - E_{X_1^{\varphi}} [h (X_2 + N_0)] = h(Y^{\varphi}) - h (X_2 + N_0),$$

where $h (X_2 + N_0)$ is a constant, and step (a) roots in the translation invariance of entropy.

Appendix 2

First of all, we note that $h(Y)$ is well-defined and uniformly continuous. Then we use the random variable H_i to represent the entries in the channel matrix $\mathbf{H}_i$, and the phase response of H_i is assumed to be uniformly distributed in $[0, 2\pi]$. Thus

$e^{j\varphi_1} H_i$ and $e^{j\varphi_2} H_i$ also follow the same distribution as H_i with any φ_1 and φ_2. When $r \to \infty$, even with different rotation angles, the empirical distribution of elements in $\mathbf{H}_i$ are the same due to asymptotic equipartition property. Therefore, with the uniform continuity of $h(Y)$, $h(Y^{\varphi_1})$ approaches $h(Y^{\varphi_2})$ with probability one.

References

1. Z. Ding et al., Application of non-orthogonal multiple access in LTE and 5G networks. IEEE Commun. Mag. **55**(2), 185–191 (2017)
2. J. Harshan, B.S. Rajan, On two-user Gaussian multiple access channels With finite input constellations. IEEE Trans. Inf. Theory **57**(3), 1299–1327 (2011)
3. S. Kundu, B.S. Rajan, Adaptive constellation rotation scheme for two-user fading MAC with quantized fade state feedback. IEEE Trans. Wirel. Commun. **12**(3), 1073–1083 (2013)
4. X. Xiao, et al., Joint optimization scheme and sum constellation distribution for multi-user Gaussian multiple access channels with finite input constellations, in *Proceedings of the AusCTW* (IEEE, Melbourne, Australia, 2016), pp. 130–135
5. M.J. Hagh, M.R. Soleymani, Constellation rotation for DVB multiple access channels With Raptor coding. IEEE Trans. Broadcast. **59**(2), 290–297 (2013)
6. J. Hershey, et al., Approximating the KL divergence between Gaussian mixture models, in *Proceedings of the IEEE ICASSP* (IEEE, Honolulu, USA, 2007), pp. 317–320
7. Discussion on the feasibility of advanced MU-detector, *R1–166098, 3GPP RAN1 #86* (Huawei HiSilicon, Gothenburg, Sweden, 2016)
8. S. Chen et al., Pattern division multiple access − A novel nonorthogonal multiple access for fifth-generation radio networks. IEEE Trans. Veh. Tech. **66**(4), 3185–3196 (2017)
9. Y. Wu et al., Linear precoding for MIMO broadcast channels With finite-alphabet constraints. IEEE Trans. Wirel. Commun. **11**(8), 2906–2920 (2012)
10. M. Cheng, Y. Wu, Y. Chen, Capacity analysis for non-orthogonal overloading transmissions under constellation constraints, in *Proceedings of the IEEE WCSP* (IEEE, Nanjing, China, 2015), pp. 1–5
11. Z. Ding, et al., On the design of MIMO-NOMA downlink and uplink transmission, in *Proceedings of the IEEE ICC* (IEEE, Kuala Lumpur, Malaysia, 2016), pp. 1–6

Chapter 5
Rate-Adaptive Design for Multiple Access

Abstract In this chapter, we discuss the rate-adaptive design for multiple access. Section 5.1 introduces the motivation of applying rate-adaptive multiple access (RAMA) for uplink grant-free transmission. Section 5.2 describes the system and channel model of grant-free transmission. Section 5.3 introduces the proposed RAMA scheme and some implementation issues. Section 5.4 conducts the theoretical analysis on conv-GF and RAMA. Section 5.5 provides two design methods on RAMA amenable constellations. Section 5.6 presents the simulation results. Section 5.7 illustrates the conclusions.

5.1 Introduction

The next generation wireless communication network (5G) is expected to support various diversified usage scenarios with different performance requirements. Specifically, the most important usage scenarios for radio access are categorized into three families by Third Generation Partnership Project (3GPP): enhanced mobile broadband (eMBB), ultra-reliable and low latency communication (uRLLC), and massive machine type communication (mMTC) [1]. While eMBB aims at offering higher peak data rates and higher system throughput in mobile hotspots, the rest two scenarios focus on the machine type communications, where mMTC is about serving massive devices with small and sporadic packets, and URLLC addresses the applications with very rigorous requirements on latency and reliability. Accordingly, the machine type communications are the extended use cases for 5G, compared with the current 4G system.

Currently, a scheduling request & grant-based access mechanism is employed in uplink data transmission. However, as shown in [2], the grant-based access mechanism expends tens of milliseconds on the signaling intersection, and the signaling overhead ratio approaches nearly a half with small packets (e.g. packets which contain dozens of bytes). Therefore, the latency and overhead requirements cannot be satisfied with grant-based access mechanism.

Recently, the grant-free access mechanism has attracted much attention from both industry and academia [3, 4], where the signaling procedure is greatly simplified. In uplink grant-free access, once the users have data in buffer, they instantly transmit their signals on the pre-configured physical resources instead of waiting for grant signaling. Thus, the grant-free access mechanism is especially suitable for uRLLC and mMTC since it reduces the latency and signaling overhead by avoiding complicated signaling intersections between users and the base station (BS). Nevertheless, due to the decentralized access, unexpected collisions may occur in grant-free access and may deteriorate the system performance. As a result, the prospects and challenges of grant-free access have motivated the researchers to make further investigations on the grant-free transceivers to fulfill the requirements of machine type communications in 5G.

5.1.1   Related Work and Motivation

During the past several decades, there have already been two research directions towards the problem of multi-user access in wireless networks, i.e., network oriented research and Shannon theory oriented research.

In the former research direction, packet transition channel model is assumed, and the researchers mainly focus on the protocol design of media access control layer. One typical example is the slotted ALOHA (SA) protocol proposed for network communications [5]. After that, the listening and backoff based SA is adopted in WIFI system [6]. In more recent years, a class of graphical based SA schemes have been proposed [7], where the random packet transmission is described by a Tanner graph.

In the latter research direction, multi-user information theory is exploited to design capacity achieving multiple access technologies. In 1970s, the capacity region of uplink multiple access channel (MAC) is derived, known as the Cover-Wyner region [8]. To approach the corner points or the hypotenuse of the Cover-Wyner region, successive interference cancellation (SIC) detection or maximum likelihood (ML) detection should be deployed at the receiver. However, these advanced receivers are not widely accepted by the industry and the academia, due to the concern of high computational complexity, until half a century of Moore's law has made the complexity less noticeable. In the most recent decade, non-orthogonal multiple access (NOMA) technologies, which is promising to achieve the entire Cover-Wyner region, have attracted much attention. By multiplexing different users' signals on the same physical resource, and employing advanced detector at the receiver, NOMA possesses higher spectral efficiency and higher overloading compared to the conventional orthogonal multiple access (OMA).

However, none of the above researches can fully describe the grant-free transmission, as pointed out in [9, 10]. We compares the grant-free access model with the access models dealt with in the above two research directions in Fig. 5.1. First of all, a two-user grant-free access model is presented in Fig. 5.1a, where each user activates according to a random variable S_i, $i = 1, 2$, and the received signal is polluted by the noise. In network-oriented research, the grant-free access channel is modeled with

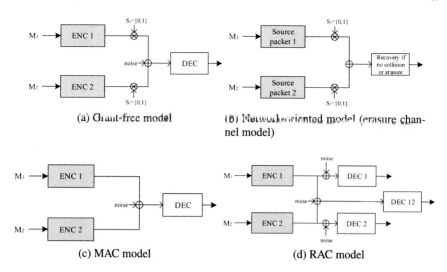

Fig. 5.1 Illustrations of MAC, grant-free transmission, and RAC models. M_1 and M_2 are the messages to be transmitted

erasure channel model, as shown in Fig. 5.1b, where the discontinuous packet arrival can be described, but the underlaid physical layer procedures are ignored, i.e., the noise and the potential near-far effect between two users cannot be modeled. Similarly, the grant-free access cannot be modeled by the MAC model either, as shown in Fig. 5.1c, since the MAC model assumes full buffer traffic and neglects the random packet arrivals.

Efforts have been made to provide a universal description of grant-free access by bridging the gap between these two research directions. A two-user random access channel (RAC) model is proposed in [10], as presented in Fig. 5.1d, where three auxiliary receivers are introduced to represent three different user activation conditions, i.e., user-1 activates, user-2 activates and both users activate. The RAC model roots in the Shannon information theory, while it also reflects the burst transmissions of grant-free access. Since the RAC model involves multiple transmitters and multiple receivers, it is similar to the interference channel (IC) model proposed in multi-user information theory [11]. The capacity region of RAC is defined by taking the closure of the unions of achievable rate tuples at user-1 and user-2 with respects to above receivers, which is a classical information theoretic approach. Obviously, different from in the MAC model, where the NOMA technologies can approach the entire capacity region of MAC, they are not capacity achieving in RAC. As illustrated in [10], rate splitting technique is required to approach the Shannon limit, at least in two-user RAC.

However, the existing literatures mainly focus on very theoretical cases, and do not provide practical designs to incorporate the rate-splitting in grant-free transmission. Also they usually assume a grant-free access system with up to two users, which is far from the requirements of 5G. And the advantages are not clarified when the number of users is more than two. Nevertheless, they do provide with the insightful hint, that is to use rate splitting for grant-free users. In this chapter, we aim to design a practical multiple access scheme to address the unpredictable interference in grant-free access, and to fully utilize the underlaid physical channel. In the meantime, the performance gain of the proposed scheme over conventional grant-free access (conv-GF) is also clarified.

5.1.2   Contributions

5.1.2.1   Novel Grant-Free Access Scheme

To combat the unpredictable interference in grant-free access, we propose a novel multiple access scheme, namely rate-adaptive multiple access (RAMA). The main idea of RAMA is to incorporate rate splitting at the transmitter, and employ SIC receiving at the receiver. With rate splitting, the total transmission power is unequally split into several layers and each layer is assigned with an independent codeword, where the multiple layers of a single user hold unequal protection property (UEP). The layers with UEP can be successfully recovered under different levels of interference, i.e., when the collision is high, the layers with high priority can be recovered, while when collision is low, the layers with low priority can take the advantage of the channel. Besides, SIC receiver is employed to mitigate the interference among the layers and the users. In this way, RAMA can enhance the system throughput, while reduce the outage probability simultaneously. Although the grant-free users cannot know the channel conditions in advance, the actual achievable transmission rates can still adapt to the real-time conditions of channel. Thus RAMA actually achieves the rate adaptation, which is similar to the link adaption in the grant-based transmission. Also, introducing more layers at the transmitter can provide more opportunities for interference cancellation which makes RAMA more robust than conv-GF with high user activation probability.

5.1.2.2   Clarification on Performance Gain

Another contribution of our work is that we analytically clarify the performance gain of RAMA over conv-GF. The existing literatures have shown that incorporating rate splitting in grant-free access with two users in an information theoretical approach can achieve performance gain. However, it is hard to illustrate the gain with multiple users, since the information theoretical model of grant-free is rather complicated with more than two transmitters. In this chapter, the throughput performance and the outage probability are analyzed with statistical methods, where users are randomly deployed in the cell. We then formulate the exact expressions of the outage probability and throughput of RAMA and conv-GF. Analysis and simulation results reveal that RAMA can achieve higher sum throughput as well as lower outage probability which illustrates that the fairness among users are also improved.

5.1.2.3 Constellation Design

To facilitate the proposed RAMA scheme, we propose two RAMA amenable constellation design methods, namely overlapping method and bundling method, respectively. Specially, the constellations, designed by the overlapping method, are composed of several base constellations, where the parameter optimization methods are discussed, including the optimizations of power coefficients and relative rotation angles among the base constellations.

5.2 System Model

Consider a grant-free access network as shown in Fig. 5.2. Suppose that a total of M users are randomly distributed in the cell with the maximum distance R_1 and the minimum distance R_2, and the BS is deployed in the center of the cell. We model packet arrivals at the user side as Poisson distribution, and define Poisson arrival rate as γ. In the network, the users are always in inactive mode to save energy if their buffer is empty. Once there are packets in the buffer, the users transfer to active mode and instantly transmit the data packets without grant signaling from the BS. Without loss of generality, we suppose that the BS has full knowledge of user activation information, e.g. via user-specific preamble or simplified random access procedure [10, 12], while the users have no knowledge about other activated users. Note that in this chapter we focus on the one-shot grant-free transmission [13], and do not consider the retransmission or hybrid automatic repeat request (HARQ). At each time index-t, one physical resource block is defined for grant-free access. Define $b_m^t \in \{0, 1\}^k$ as the information bit sequence of an active user-m ($1 \leq m \leq M$) with length-k at time index-t. In conv-GF, b_m^t is first encoded into a single coded bit sequence $c_m^t \in \{0, 1\}^n$ with coding rate r_m^{conv}, and then modulated into a complex symbol sequence $x_m^t \in \{\mathcal{X}\}^{n/\log_2(|\mathcal{X}|)}$, where $\mathcal{X}$ is the constellation with cardinality $|\mathcal{X}|$. Each symbol sequence occupies the entire physical resource block. Furthermore, we assume Rayleigh block fading channel model, where the Rayleigh distributed small-scale fading coefficient remains a constant within each block, and the fading is independent and identically distributed (i.i.d.) among different blocks or users. Therefore the channel coefficient between the BS and user-m at time index-t is given by $h_m^t = \dfrac{g_m^t}{\sqrt{d_m^\alpha}}$ [14], where g_m^t is Rayleigh fading coefficient, d_m is the distance between the BS and the user-m, and α is the decay exponent. Without loss of generality, we assume $\alpha = 2$ [15], and assume the transmission power of each

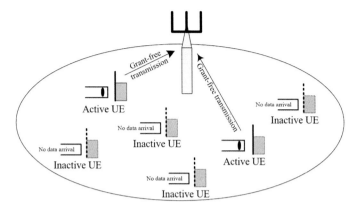

Fig. 5.2 Grant-free access system model

user is P. Therefore, the received signal at the BS at time index-t is formulated as follows

$$\mathbf{y}^t = \sum_{m=1}^{M} I_m^t h_m^t \sqrt{P} \mathbf{x}_m^t + \mathbf{n}^t, \ I_m^t = \begin{cases} 1, & \text{active} \\ 0, & \text{inactive} \end{cases}, \tag{5.1}$$

where I_m^t indicates the user activation, and n^t is the additive white Gaussian noise with zero mean and variance σ^2. At the receiver, advanced multi-user detector (MUD) is usually employed to mitigate the mutual interference among the users and to distinguish different users' data streams. For simplification, we define T^t as the normalized throughput of the network which equals to the number of successfully transmitted packets at time index-t, and the average normalized throughput is given by $T = \mathrm{E}_t\{T^t\}$. The outage is defined as the event that a packet is not successfully decoded in given time index.

In this chapter, we assume that the BS deploys the SIC receiver, which is a typical MUD in NOMA [4], to accomplish a good tradeoff between the detection accuracy and the computational complexity. The main idea of SIC is to firstly recover and cancel the data streams with high priority while regarding the other signals as noise, and then take advantages of the residual signals. In the conventional grant-based NOMA, the users are scheduled by the BS to deliberately and cautiously multiplex together to ensure low detection error probability. However, as one may expect, the random superposition of signals in grant-free data transmission may not facilitate SIC receiving. Therefore, more elaborate designs should be made at the transmitters to enhance the total throughput and reduce the outage probability of grant-free access system.

5.3 Rate-Adaptive Multiple Access

In grant-based access, each user transmits a codeword in a slot with a certain coding rate derived by channel estimation. However, in grant-free access, the users cannot anticipate the real-time traffic load and may experience unexpected interference from other active users. We illustrate the achievable data rates versus the interference with different grant-free access schemes in Fig. 5.3. Conv-GF adopts the same transmission strategy as in grant-based access, which may lead to performance loss compared to the theoretical limit, as shown by the red arrows in Fig. 5.3a. When the interference is lower than the threshold, the user cannot fully utilize the potential of channel, and when the interference is higher than the threshold, the data cannot even be successfully recovered. One may expect an ideal grant-free access scheme where the achievable data rate at receiver can automatically adapt to the interference and thus follow the theoretical limits, as shown in Fig. 5.3b. However, this is not realistic. In this section, we propose a rate-adaptive multiple access (RAMA) scheme for grant-free data transmission, which is based on the rate-splitting technique, and can be regarded as the an approximation to the ideal grant-free access as illustrated in Fig. 5.3c. With this aim, we firstly introduce the rate-splitting technique before presenting the proposed RAMA scheme.

5.3.1 Rate-Splitting Principle

Rate-splitting (RS) was originally introduced in the multi-user information theory as a technique to prove the capacity bounds of broadcasting channel (BC), multiple access channel (MAC), and interference channel (IC) [16]. The core idea of RS is to split the original message into two or several independent layers, and transmit them simultaneously. During these years, RS has attracted the attention of researchers for its potentials to reach every points in MAC [17], to enhance the fairness among the users in the network [18], and to promote the security in MIMO network [19], etc.

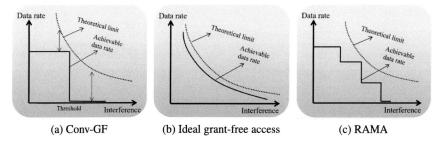

(a) Conv-GF   (b) Ideal grant-free access   (c) RAMA

Fig. 5.3 Achievable data rate vs interference, with different grant-free access schemes

5.3.2   RAMA for Grant-Free Transmission

The proposed RAMA scheme is demonstrated in Fig. 5.4, where RS technique and SIC are adopted at the transmitters and the receiver respectively. When each user has data in buffer, it instantly transmits signals according to the RAMA scheme, as shown in Fig. 5.4a, with three steps.

- Step 1: data reorganization. At the active user-m, information bit sequence $b_m^t \in \{0, 1\}^n$ is partitioned and reorganized by a bijection B

$$\text{B} : b_m^t \mapsto (\beta_{m,1}^t, \beta_{m,2}^t \cdots \beta_{m,L_m}^t),$$

where L_m is the number of layers, and $\beta_{m,2}^t (1 \le l \le L_m)$ is the information bit sequence for lth layer. We assign different priority levels to data layers, where layers with higher priority will experience greater protection.
- Step 2: single-layer channel coding. Each sub-sequence $\beta_{m,l}^t$ is encoded with a channel encoder ENC_l with rate $r_{m,l}^{\text{RAMA}}$

$$\text{ENC}_l : \beta_{m,l}^t \mapsto c_{m,l}^t,$$

where $c_{m,l}^t$ is the coded bit sequence. Generally, high priority layers are encoded with low coding rate.

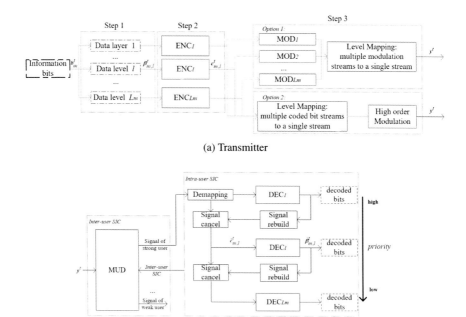

(a) Transmitter

(b) Receiver

Fig. 5.4 RAMA transmitter and receiver structure

- Step 3: layer-aggregation. Two options can be used to aggregate different layers into one symbol sequence.
In option 1, each coding layer $c_{m,l}^t$ is independently modulated with modulator MOD_l

$$\text{MOD}_l : c_{m,l}^t \mapsto x_{m,l}^t,$$

where $x_{m,l}^t$ is the modulation symbol sequence with each element drawn from the constellation $\mathcal{X}_{m,l}$. All layers $x_{m,l}^t (1 \leq l \leq L_m)$ are then superimposed together to get the composite constellation symbol sequence x_m^t with certain power coefficient $\boldsymbol{\lambda}_m^t = [\lambda_{m,1}^t, \lambda_{m,2}^t \dots \lambda_{m,L_m}^t]$ and phase rotation angle $\boldsymbol{\vartheta}_m^t = [\vartheta_{m,1}^t, \vartheta_{m,2}^t \dots \vartheta_{m,L_m}^t]$, where x_m^t is given by

$$x_m^t = \sum_{l=1}^{L_m} \lambda_{m,l}^t x_{m,l}^t e^{j\vartheta_{m,l}^t},$$

$$x_m^t \in \{\mathcal{X}_{(\lambda_m^t, \vartheta_m^t)}\}^{n/\log_2\left(\left|\mathcal{X}_{(\lambda_m^t, \vartheta_m^t)}\right|\right)}, \tag{5.2}$$

and $\mathcal{X}_{(\lambda_m^t, \vartheta_m^t)}$ is the composite constellation defined by $\boldsymbol{\lambda}_m^t$ and $\boldsymbol{\vartheta}_m^t$. The layers with higher priority are assigned with larger power coefficients.
In option 2, multiple coding layers are firstly mapped to a single bit stream and then modulated with high-order constellation, similar to the coded modulation, as $(c_{m,1}^t, c_{m,2}^t \cdots c_{m,L_m}^t) \mapsto x_m^t$, where $x_m^t \in \{\mathcal{X}\}^{n/\log_2(|\mathcal{X}|)}$. The mapping ensures that the coded bits of higher priority layers holds larger minimum Euclidean distances.

At the receiver, multi-user detection algorithm should be employed to distinguish different users' signals, since multiple users may collide due to the uncoordinated transmission. We show an optimal SIC-based detection algorithm in Algorithm 5.1, which requires to traverse all possible SIC orders. To reduce the computational complexity, we propose a simplified detection algorithm as demonstrated Algorithm 5.2. Note that, by employing Algorithm 5.2, we can simplify the analysis of outage performance in Sect. 5.5.

As shown in Fig. 5.3c, The advantage of RAMA can be intuitively illustrated as follows. In RAMA, the data of each user is transmitted with multiple signal layers, where all the layers have different reliability. As for a certain user, when the external interference is significant, only the signal layer with higher reliability can be solved.

Algorithm 5.1 Optimal SIC-Based Detection Algorithm

Require: the received signal y^t.
Ensure: the estimated information bits.
1: Transverse all possible SIC orders and conduct SIC receiving for each SIC order.
2: Find the optimal SIC order which achieves the largest throughput and output the estimated information bits.

Algorithm 5.2 The Proposed SIC-Based Detection Algorithm

Require: the received signal y^t, the number of active user M_{ac}, maximum SIC iteration number
  S_{max}, maximum layer number L_{max}, and flag f.
Ensure: the estimated information bits.
1: Set $f = 1$.
2: **while** $f = 1$ & $1 \leq s \leq S_{max}$ **do**
3:    Set $f = 0$.
4:    **for** $1 \leq l \leq L_{max}$ **do**
5:      **for** $1 \leq m \leq M_{ac}$ **do**
6:          [Step 1]: Detect the lth signal layer of the user with mth strongest channel gain while
            regarding interference as noise.
7:          **if** this signal layer is successfully recovered **then**
8:              [Step 2]: Output the estimated information bits of this signal layer.
9:              [Step 3]: Reconstruct and cancel this signal layer from y^t.
10:             [Step 4]: Set flag $f = 1$.
11:          **end if**
12:      **end for**
13:    **end for**
14: **end while**

Otherwise, when the external interference is not significant, the signal layers with lower reliability can also be successfully recovered. Besides, the layered structure can also mitigate the mutual interference, since the recovered signal layers can be reconstructed and canceled via SIC receiving. As a result, even if the grant-free user cannot foresee the channel occupancy, each user's actual transmission rate can still adapt to the real-time conditions of channel.

5.3.3 Implementation Issues

5.3.3.1 Priority Setting

As mentioned above, the multiple layers in the transmission signals of RAMA exhibits UEP, and the data with disparate priority shall be mapped to corresponding layers. Therefore, the order of the importances of data sets shall be decided in practice. The data sets can be randomly assigned with different priority. Moreover, in some usage scenarios, different data sets naturally have different levels of importance. For example, in mMTC, data sets may contain user identity and application data. The data set containing the user identity is regarded as having high priority. Once the BS knows the user identities, the BS may schedule these users with grant-based transmission to mitigate the collision [20]. Another example happens when grant-free uplink transmission collides with the uplink control information (UCI) on the same resources [21]. In this case, the user may piggyback the UCI report into grant-free data transmission, and the UCI report and grant-free data has different priority, e.g. if the data is for URLLC and the UCI is for eMBB, the former has higher priority.

5.3.3.2 Frame Structure

The proposed RAMA scheme can be incorporated into existing frame structure designed for grant-free transmission [23]. However, the transmission block sizes (TBSs) defined for LTE may not satisfy the need of RAMA, since RAMA contains more than one data blocks in each transmission and some data blocks may only have much less amount of bits. Thus more TBSs should be defined for RAMA.

5.3.3.3 Retransmission

Due to the UEP of RAMA, some signal layers may not have enough signal to interference and noise ratio (SINR) to be recovered, and retransmission can be employed to make use of the received signals. For the retransmission, either grant-based or grant-free transmission is available depending on specific reliability or latency requirements.

5.4 Performance Analysis of Conv-GF and RAMA

In this section, we analyze and compare the outage and throughput performance of both conv-GF and RAMA, and show the advantages of RAMA.

5.4.1 Outage Performance Analysis of Grant-Free Access

The performance of conv-GF and RAMA is analyzed in incremental steps. First of all, we study the channel statistics in the following *Lemma* 5.1.

Lemma 5.1 *For each active user, the probability density function (PDF) of channel gains at time index-t, i.e., $|h^t|^2$, is given by*

$$
f_{|h^t|^2}(z) = -\frac{1}{\left(R_1^2 - R_2^2\right) z^2} \left(\left(e^{-R_1^2 z/2} \left(R_1^2 z + 2 \right) \right) \right.
$$
$$
\left. - \left(e^{-R_2^2 z/2} \left(R_2^2 z + 2 \right) \right) \right),
$$

(5.3)

and the the cumulative density function (CDF) of $|h_m^t|^2$ is

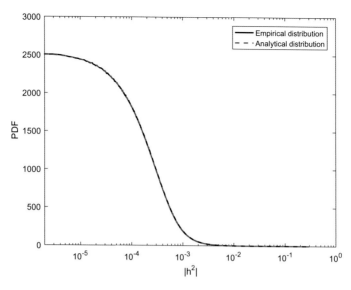

Fig. 5.5 Comparison between empirical and analytical distributions of $|h^t|^2$, with $R_1 = 100$, $R_2 = 10$, and 10^8 samples

$$F_{|h^t|^2}(z) = \int_{x=0}^{z} f_{|h^t_m|^2}(x)\,dx$$

$$= 1 + \frac{2e^{-(R_1^2 z)/2} - 2e^{-(R_2^2 z)/2}}{z\left(R_1^2 - R_2^2\right)}. \tag{5.4}$$

Proof Please refer to Appendix 1.   □

The empirical and analytical distributions of $|h^t|^2$ are compared in Fig. 5.5, which shows a perfect match. In the following, we omit the index t for simplicity.

Denote the event that M_{ac} users become active at time index-t as $K_{M_{\mathrm{ac}}}$, and its probability can be given by

$$P(K_{M_{\mathrm{ac}}}) = \frac{(\lambda M)^{M_{\mathrm{ac}}} e^{-\lambda M}}{M_{\mathrm{ac}}!}. \tag{5.5}$$

Further, define $G_{\{|h_1|^2, \cdots |h_{M_{\mathrm{ac}}}|^2\}}$ as the event where the channel gains of the M_{ac} active users form the set $\mathbf{H} = \{|h_1|^2, \cdots |h_{M_{\mathrm{ac}}}|^2\}$ at time index-t. Note that $G_{\{\cdots|h_i|^2, \cdots|h_j|^2\cdots\}}$ and $G_{\{\cdots|h_j|^2, \cdots|h_i|^2\cdots\}}$ are exactly the same event.

Proposition 5.1 *The PDF of* $G_{\{|h_1|^2, \cdots|h_{M_{ac}}|^2\}}$ *is given by*

$$P\left(G_{\{|h_1|^2, \cdots|h_{M_{ac}}|^2\}}\right) = M_{ac}! \prod_{m=1}^{M_{ac}} f_{|h|^2}(|h_m|^2). \tag{5.6}$$

Proof For an active user, the PDF of $|h_m|^2$ is $f_{|h|^2}(|h_m|^2)$. Since the channel coefficients of different users are independent, the joint probability density of channel gain vector $\left(|h_1|^2, |h_2|^2, \cdots |h_m|^2\right)$ is $\prod_{m=1}^{M_{ac}} f_{|h|^2}(|h_m|^2)$. We note that sweeping the elements in **H** does not make it a distinct event, and there are a total number of $M_{ac}!$ permutations of $\left(|h_1|^2, |h_2|^2, \cdots |h_m|^2\right)$, which corresponds to the same event $G_{\{|h_1|^2, \cdots |h_{M_{ac}}|^2\}}$. Thus the PDF of $G_{\{|h_1|^2, \cdots |h_{M_{ac}}|^2\}}$ is given by multiplexing $M_{ac}!$ with $\prod_{m=1}^{M_{ac}} f_{|h|^2}(|h_m|^2)$. $\square$

Up to now, the channel gains of M_{ac} users are unordered, and thus is hard to analyze with SIC receiver. However, since permuting the elements in **H** does not change the set itself, we can always assumed that the set $\{|h_1|^2, \cdots |h_{M_{ac}}|^2\}$ is sorted with $|h_i| < |h_j|$ if $i > j$. The normalization condition of P_G holds as follows

$$\int_{z_1 \geq \cdots \geq z_{M_{ac}} \geq 0} P\left(G_{\{z_1, \cdots z_{M_{ac}}\}}\right) dz_1 \ldots z_{M_{ac}} = 1. \tag{5.7}$$

Few literatures have considered the outage probability of uplink NOMA. Zhang et al. [14] studied the outage probability of NOMA, where the outage events of the users in the uplink NOMA system are considered to be mutually independent, and the users which is successfully recovered is considered as having no correlation with the remaining users. However, this argument is incomplete, since the outage event of the previous users may indicate the channel conditions of the remaining users. In the following example, we show the argument in [14] is incomplete.

Example 5.1 Consider a special two-user uplink NOMA system with unit transmission power and unit-variance Gaussian noise. The ordered channel coefficients, namely h_1 and h_2 ($h_1 > h_2$), may choose values from the set $\{2, 3\}$ with equal probability. Define the outage event of user-i as E_i, $i = 1, 2$, and E_i^c as the complementary set of E_i. Assume that the target data rates of both users are $r_1 = \log_2(1 + 1) = 1$ and $r_2 = \log_2(1 + 3) = 2$, respectively. Then we find that event E_1^c happens only when $h_1 = 3$ and $h_2 = 2$, and in this case, E_2 must happen. Otherwise, when E_1 happens, we have $h_1 = 3$ and $h_2 = 3$, or $h_1 = 2$ and $h_2 = 2$, where E_2 happens with half probability. As a result, E_1 is correlated to the E_2. In this example, we see that the outage events of previous users actually constrain the probability spaces of the channel gains of the rest users, and thus influence the outage events.

As illustrated in Example 5.1, to get the outage probability of mth user, it is not appropriate to decompose the outage event into several independent events. In stead, we directly deal with the outage event of the active users by applying high dimensional integration in the following derivations. With this aim, we define the following outage events to analyze the outage probability of conv-GF.

Without loss of generality, we assume that all users adopt the same transmission rate, i.e., $r_m^{conv} = r^{conv}$, $\forall m$ [23]. We use $E_{m,M_{ac}}^{conv}$ to represent the outage event where the signals of 1st to $(m-1)$th users are successfully recovered, and the signals of mth to M_{ac}th users cannot be recovered. In the following, we assume capacity achieving channel coding and modulation, if not specified. Thus $E_{m,M_{ac}}^{conv}$ is given by

$$E_{m,M_{ac}}^{conv} \triangleq \left\{ \hat{r}_j^{conv} \geq r_j^{conv}, 1 \leq j < m, \text{ and } \hat{r}_m^{conv} < r_m^{conv} \right\}, \tag{5.8}$$

where $\hat{r}_j^{conv} = \log_2 \left(1 + \text{SINR}_j^{conv} \right)$, and SINR_j^{conv} is the received SINR of user-m. According to (5.8), we readily have the following proposition.

Proposition 5.2 *The conditional probability of event $E_{m,M_{ac}}^{conv}$ given $G_{\{|h_1|^2, \cdots |h_{M_{ac}}|^2\}}$ is derived as,*

$$P \left(E_{m,M_{ac}}^{conv} \middle| G_{\{|h_1|^2, \cdots |h_m|^2, \cdots |h_{M_{ac}}|^2\}} \right)$$
$$= \begin{cases} 1, & C_{m,M_{ac}}^{conv} \\ 0, & \text{otherwise}, \end{cases} \quad 1 \leq m \leq M_{ac}, \tag{5.9}$$

where, $\phi = 2^{r^{conv}} - 1, 1 \geq i \geq M_{ac}$, and is given by

$$C_{m,M_{ac}}^{conv} = \left\{ (|h_1|^2, \cdots |h_{M_{ac}}|^2) \middle| \frac{|h_j|^2 P}{\sum_{i=j+1}^{M_{ac}} |h_i|^2 P + \sigma^2} \geq \phi, \right.$$
$$\left. 1 \leq j < m, \text{ and } \frac{|h_m|^2 P}{\sum_{i=m+1}^{M_{ac}} |h_i|^2 P + \sigma^2} < \phi \right\}, \tag{5.10}$$

Averaging (5.9) over the entire probability space of $G_{\{z_1, \cdots z_{M_{ac}}\}}^t$, we have

$$P \left(E_m^{conv} \right) = \int_{z_1 \geq \cdots \geq z_{M_{ac}} \geq 0} P \left(E_m^{conv} \middle| G_{\{z_1, \cdots z_{M_{ac}}\}} \right)$$
$$\times P \left(G_{\{z_1, \cdots z_{M_{ac}}\}}^t \right) dz_1 \ldots z_{M_{ac}}, \tag{5.11}$$

and the exact expression of $P \left(E_{m,M_{ac}}^{conv} \right)$ is given by (5.12).

$$P \left(E_{m,M_{ac}}^{conv} \right) = M_{ac}! \int_0^{+\infty} f_{|h|^2}(z_{M_{ac}}) \int_{z_{M_{ac}-1}=z_{M_{ac}}}^{+\infty} f_{|h|^2}(z_{M_{ac}-1})$$

$$\cdots \int_{z_{m+1}=z_{m+2}}^{+\infty} f_{|h|^2}(z_{m+1}) \int_{z_m=z_{m+1}}^{\phi(\sum_{j=m+1}^{M_{ac}} z_j + \sigma^2/P)} f_{|h|^2}(z_m)$$

$$\times \int_{z_{m-1}=\max\left(z_m, \phi(\sum_{j=m}^{M_{ac}} z_j + \frac{\sigma^2}{P})\right)}^{+\infty} f_{|h|^2}(z_{m-1}) \cdots \int_{z_1=\max\left(z_2, \phi(\sum_{j=2}^{M_{ac}} z_j + \frac{\sigma^2}{P})\right)}^{+\infty} f_{|h|^2}(z_1) dz_1 \cdots z_{M_{ac}}$$

$$\tag{5.12}$$

$$P\left(E^{conv}_{m,M_{ac}}\right) = M_{ac}! \int_{0}^{+\infty} f_{|h|^2}(z_{M_{ac}}) \int_{z_{M_{ac}-1}=z_{M_{ac}}}^{+\infty} f_{|h|^2}(z_{M_{ac}-1}) \cdots \int_{z_{m+1}=z_{m+2}}^{+\infty} f_{|h|^2}(z_{m+1})$$

$$\times \int_{z_m=z_{m+1}}^{\phi(\sum_{j=m+1}^{M_{ac}} z_j+\sigma^2/P)} f_{|h|^2}(z_m) \int_{z_{m-1}=\phi(\sum_{j=m}^{M_{ac}} z_j+\sigma^2/P)}^{+\infty} f_{|h|^2}(z_{m-1})$$

$$\cdots \int_{z_1=\phi(\sum_{j=2}^{M_{ac}} z_j+\sigma^2/P)}^{+\infty} f_{|h|^2}(z_1) dz_1 \cdots dz_m \cdots z_{M_{ac}}$$

$$(5.13)$$

Due to the non-continuous max operations in the integral regions, it is generally difficult to integrate (5.12) with either numerical intergeneration or approximation. Therefore, the exact expressions of (5.12) which do not contain the max operations should be derived. When $\phi < 1$, the exact expressions of $P\left(E^{conv}_{m,M_{ac}}\right)$ can be recursively derived as shown in Appendix 2, which does not involve the max operations. When $\phi \geq 1$, the exact expression of $P\left(E^{conv}_{m,M_{ac}}\right)$ is given by (5.13). Without loss of generality, we assume $\phi \geq 1$ in the following analysis.

However, it is still non-trivial to derive a general and closed-form expression of (5.13). However, according to the requirements in [5G traffic model], the average number of new packets in each time index is at the level of 10^0, where 2 is a typical value. Therefore, in Appendix 3, we derive the compact outage expressions of the outage probabilities for some special cases where active user number is smaller or equal to 3, i.e., $M_{ac} \leq 3$, which may constitute the mainstreams of grant-free transmission in the practice. Define the outage event of an active user with conv-GF as E^{RAMA}. Now we are ready to give the expressions of the outage and throughput performance of conv-GF.

Theorem 5.1 *The average outage probability and the throughput of conv-GF is given by*

$$P\left(E^{conv}\right) =$$

$$\frac{\sum_{M_{ac}=1}^{\infty} P\left(K_{M_{ac}}\right) \left(\sum_{m=1}^{M_{ac}} P\left(E^{conv}_{m,M_{ac}}\right)(M_{ac}-m+1)\right)}{\sum_{M_{ac}=1}^{\infty} P\left(K_{M_{ac}}\right) M_{ac}}, \qquad (5.14)$$

and

$$T^{RAMA} = \sum_{M_{ac}=1}^{\infty} \left(P\left(K_{M_{ac}}\right) \sum_{m=1}^{M_{ac}} \left(P\left(E^{conv}_{m,M_{ac}}\right)(mr^{conv})\right)\right), \qquad (5.15)$$

respectively.

Proof When M_{ac} users are active in a time index, the average amount of the outage users is $\sum_{m=1}^{M_{ac}} P\left(E^{conv}_{m,M_{ac}}\right)(M_{ac}-m+1)$. Averaging the above value over (5.5), we get (5.14). Similarly, T^{RAMA} is given by multiplexing the transmission rate of users with the successfully recovered users, as derived in (5.15).   □

With the similar approach, the outage probability of RAMA can also be derived as follows.

$$
\begin{aligned}
C^{\text{RAMA},(1)}_{\{m_1,m_2\},M_{\text{ac}}} = \Bigg\{ \Big(|h_1|^2, \cdots |h_{M_{\text{ac}}}|^2\Big) \Bigg| & \frac{|h_i|^2 \alpha P}{\sum_{j=i+1}^{M_{\text{ac}}} |h_j|^2 \alpha P + \sum_{j=1}^{M_{\text{ac}}} |h_j|^2 (1-\alpha) P + \sigma^2} \geq \varphi_1, \\
& \frac{|h_{m_1}|^2 \alpha P}{\sum_{j=m_1+1}^{M_{\text{ac}}} |h_j|^2 \alpha P + \sum_{j=1}^{M_{\text{ac}}} |h_j|^2 (1-\alpha) P + \sigma^2} < \varphi_1, \\
& \frac{|h_k|^2 (1-\alpha) P}{\sum_{j=m_1}^{M_{\text{ac}}} |h_j|^2 \alpha P + \sum_{j=k+1}^{M_{\text{ac}}} |h_j|^2 (1-\alpha) P + \sigma^2} \geq \varphi_2, \\
\text{and } & \frac{|h_{m_2}|^2 (1-\alpha) P}{\sum_{j=m_2+1}^{M_{\text{ac}}} |h_j|^2 \alpha P + \sum_{j=m_1}^{M_{\text{ac}}} |h_j|^2 (1-\alpha) P + \sigma^2} < \varphi_2, \\
& 1 \leq i < m_1, 1 \leq k < m_2, \Bigg\}
\end{aligned}
$$

$$(5.16)$$

$$
\begin{aligned}
C^{\text{RAMA},(2)}_{\{m_1,m_2\},M_{\text{ac}}} = \bigcup_{1 \leq m_{1,1} < m_1, 1 \leq m_{2,1} \leq m_2} \Bigg\{ (|h_1|^2, \cdots |h_{M_{\text{ac}}}|^2) \Big| C^{\text{RAMA},(1)}_{\{m_{1,1},m_{2,1}\},M_{\text{ac}}}, \\
\frac{|h_i|^2 \alpha P}{\sum_{j=i+1}^{M_{\text{ac}}} |h_j|^2 \alpha P + \sum_{j=m_{1,1}}^{M_{\text{ac}}} |h_j|^2 (1-\alpha) P + \sigma^2} \geq \varphi_1, \\
\frac{|h_{m_1}|^2 \alpha P}{\sum_{j=m_1+1}^{M_{\text{ac}}} |h_j|^2 \alpha P + \sum_{j=m_{1,1}}^{M_{\text{ac}}} |h_j|^2 (1-\alpha) P + \sigma^2} < \varphi_1, \\
\frac{(m_2-m_{2,1})|h_k|^2 (1-\alpha) P}{\sum_{j=m_1}^{M_{\text{ac}}} |h_j|^2 \alpha P + \sum_{j=k+1}^{M_{\text{ac}}} |h_j|^2 (1-\alpha) P + \sigma^2} \geq (m_2-m_{2,1})\varphi_2, \\
\text{and } \frac{|h_{m_2}|^2 (1-\alpha) P}{\sum_{j=m_2+1}^{M_{\text{ac}}} |h_j|^2 \alpha P + \sum_{j=m_1}^{M_{\text{ac}}} |h_j|^2 (1-\alpha) P + \sigma^2} < \varphi_2, \\
m_{1,1} \leq i < m_1, m_{2,1} \leq k \leq m_2 \Bigg\}.
\end{aligned}
$$

$$(5.17)$$

5.4.2 Outage Performance Analysis of RAMA

In this subsection, we analyze the outage performance of RAMA. Without loss of generality, we assume that each user splits its signal into two layers with RAMA, since introducing more layers may lead to severe error propagation [24]. Besides, we assume all users adopt the same transmission procedure with the same coefficients, and denote r_1^{RAMA} and r_2^{RAMA} as the transmission rate of the layer-1 and the layer-2 at each user, respectively. The power splitting ratio is defined as α for each user, i.e., the transmission power of the layer-1 and the layer-2 are αP and $(1 - \alpha)P$, respectively.

Similar to conv-GF, we denote $E_{\{m_1,m_2\},M_{\text{ac}}}^{\text{RAMA}}$ as the outage event where the 1st to $(m_1 - 1)$th users's first layers and the 1st to $(m_2 - 1)$th users's second layers are successfully recovered, respectively, while the rest of the layers cannot be recovered. Furthermore, we assume that the layer-1 exhibits higher protection than the layer-2, i.e., the layer-1 can always be successfully detected once the layer-2 can be successfully detected. Therefore, $E_{\{m_1,m_2\},M_{\text{ac}}}^{\text{RAMA}}$ is given by

$$E_{\{m_1,m_2\},M_{\text{ac}}}^{\text{RAMA}} \triangleq \left\{ \hat{r}_{j,1}^{\text{RAMA}} \geq r_1^{\text{RAMA}}, 1 \leq j < m_1, \right.$$
$$\hat{r}_{k,2}^{\text{RAMA}} \geq r_2^{\text{RAMA}}, 1 \leq k < m_2, \qquad (5.18)$$
$$\left. \hat{r}_{m_1,1}^{\text{RAMA}} < r_1^{\text{RAMA}}, \text{ and } \hat{r}_{m_2,2}^{\text{RAMA}} < r_2^{\text{RAMA}} \right\}.$$

where $\hat{r}_{j,l}^{\text{RAMA}} = \log_2\left(1 + \text{SINR}_{j,l}^{\text{RAMA}}\right)$, and $\text{SINR}_{j,l}^{\text{RAMA}}$ is the received SINR of the lth layer of the jth user. Due to the fact that the 1st layer exhibit higher protection than the 2nd layer, we always have $m_1 \geq m_2$. Besides, we define the outage event of active users, namely E_m^{RAMA}, which happens when at least 1 layer of the 1st to $(m - 1)$th users are recovered, and none of the layers of mth to M_{ac}th users is successfully decoded, and E_m^{RAMA} can be readily defined as

$$E_m^{\text{RAMA}} \triangleq \bigcup E_{\{m,m_2\},M_{\text{ac}}}^{\text{RAMA}}, 1 \leq m_2 \leq m. \qquad (5.19)$$

As aforementioned, Algorithm 5.1 is the optimal SIC-based multi-user detection algorithm for RAMA. However, since Algorithm 5.1 involves a traversing operation, which does not have an exact mathematical expression, we study the performance of RAMA by assuming Algorithm 5.2. Before that, we first show the optimality of the proposed Algorithm 5.2 by Lemma 5.2.

Lemma 5.2 *Assume that all users split and encode their signals with the same power coefficients and the same transmission rates, respectively, the proposed Algorithm 5.1 is the optimal SIC-based multi-user detection algorithm of RAMA, i.e., Algorithm 5.1 achieves the same outage and throughput performance as Algorithm 5.2*

Proof First of all, we note that Algorithm 5.1 traverse all possible successive cancellation orders, and therefore it is the optimal multi-user detection algrithm based on SIC. Next, we show the optimality of Algorithm 5.2 by contradiction. Assume that the lth layer of the mth user happens to be decoded by Algorithm 5.1 but not by Algorithm 5.2. According to the assumptions of this chapter, the 1st to lth layers of the 1st to $(m - 1)$th users, and the 1st to $(l - 1)$th layers of the mth user can be successfully recovered by both Algorithms 5.1 and 5.2. After canceling the aforementioned layers, the lth layer of mth users is the most reliable layer among the remaining signal layers. Therefore, Algorithm 5.1 will recover this layer while regarding other layers as noise, according to the assumption. However, Algorithm 5.2 can also decode this layer just as Algorithm 5.1, which contradicts the assumption. After all, Algorithm 5.2 is optimal. □

To model the effect of SIC receiving, we define $E^{RAMA,(s)}_{\{m_1,m_2\},M_{ac}}$ as the outage event that, after sth iteration in Algorithm 5.2, m_1th user's first layer and m_2th user's second layer cannot be successfully decoded, while the 1th to $(m_1 - 1)$th user's first layers and the 1th to $(m_2 - 1)$th user's second layers, when M_{ac} users are active. In this case, $m_1 \geq m_2$.

Proposition 5.3 *The conditional probability of event* $E^{RAMA,1}_{\{m_1,m_2\},M_{ac}}$ *given* $G_{\{|h_1|^2,\cdots|h_{M_{ac}}|^2\}}$ *is derived as*

$$
P\left(E^{RAMA,(1)}_{\{m_1,m_2\},M_{ac}}\middle| G_{\{|h_1|^2,\cdots|h_{M_{ac}}|^2\}}\right)
$$

$$
= \begin{cases} 1, & \{(|h_1|^2,\cdots|h_{M_{ac}}|^2)\in C^{RAMA,(1)}_{\{m_1,m_2\},M_{ac}}\}, \\ 0, & otherwise, \end{cases} \tag{5.20}
$$

$$
1 \leq m_1 \leq M_{ac}, 1 \leq m_2 \leq M_{ac},
$$

and $C^{RAMA,(1)}_{\{m_1,m_2\},M_{ac}}$ *is a region given by (5.16), where* $\varphi_i = 2^{r_i^{RAMA}} - 1, 1 \leq i \leq 2$.
   Similarly, $P\left(E^{RAMA,(2)}_{\{m_1,m_2\},M_{ac}}\middle| G_{\{|h_1|^2,\cdots|h_m|^2,\cdots|h_{M_{ac}}|^2\}}\right)$ *is given by*

$$
P\left(E^{RAMA,(2)}_{\{m_1,m_2\},M_{ac}}\middle| G_{\{|h_1|^2,\cdots|h_m|^2,\cdots|h_{M_{ac}}|^2\}}\right)
$$

$$
= \begin{cases} 1, & \{(|h_1|^2,\cdots|h_{M_{ac}}|^2)\in C^{RAMA,(2)}_{\{m_1,m_2\},M_{ac}}\}, \\ 0, & otherwise, \end{cases} \tag{5.21}
$$

$$
1 \leq m_1 \leq M_{ac}, 1 \leq m_2 \leq M_{ac},
$$

where, $C^{RAMA,(2)}_{\{m_1,m_2\},M_{ac}}$ *is a region given by (5.17).*
   Using mathematical induction, the general expression of
$P\left(E^{RAMA,(s+1)}_{\{m_1,m_2\},M_{ac}}\middle| G_{\{|h_1|^2,\cdots|h_m|^2,\cdots|h_{M_{ac}}|^2\}}\right)$ *is given by*

$$
P\left(E^{RAMA,(s+1)}_{\{m_1,m_2\},M_{ac}}\middle| G_{\{|h_1|^2,\cdots|h_m|^2,\cdots|h_{M_{ac}}|^2\}}\right)
$$

$$
= \begin{cases} 1, & \{(|h_1|^2,\cdots|h_{M_{ac}}|^2)\in C^{RAMA,(s+1)}_{\{m_1,m_2\},M_{ac}}\}, \\ 0, & otherwise, \end{cases} \tag{5.22}
$$

$$
1 \leq m_1 \leq M_{ac}, 1 \leq m_2 \leq M_{ac},
$$

and $C^{RAMA,(s+1)}_{\{m_1,m_2\},M_{ac}}$ *is a region given by (5.23).*

$$C^{RAMA,(s+1)}_{\{m_1,m_2\},M_{ac}} = \bigcup_{1\leq m_{1,s}<m_1, 1\leq m_{2,s}\leq m_2} \left\{ (|h_1|^2,\cdots|h_{M_{ac}}|^2) \Big| C^{RAMA,(s)}_{\{m_{1,s},m_{2,s}\},M_{ac}}, \right.$$

$$\frac{|h_i|^2\alpha P}{\sum_{j=i+1}^{M_{ac}} |h_j|^2\alpha P + \sum_{j=m_{1,s}}^{M_{ac}} |h_j|^2(1-\alpha)P + \sigma^2} \geq \varphi_1,$$

$$\frac{|h_{m_1}|^2\alpha P}{\sum_{j=m_1+1}^{M_{ac}} |h_j|^2\alpha P + \sum_{j=m_{1,s}}^{M_{ac}} |h_j|^2(1-\alpha)P + \sigma^2} < \varphi_1,$$

$$\frac{(m_2-m_{2,s})|h_k|^2(1-\alpha)P}{\sum_{j=m_1}^{M_{ac}} |h_j|^2\alpha P + \sum_{j=k+1}^{M_{ac}} |h_j|^2(1-\alpha)P + \sigma^2} \geq (m_2-m_{2,s})\varphi_2,$$

$$and \quad \frac{|h_{m_2}|^2(1-\alpha)P}{\sum_{j=m_2+1}^{M_{ac}} |h_j|^2\alpha P + \sum_{j=m_1}^{M_{ac}} |h_j|^2(1-\alpha)P + \sigma^2} < \varphi_2,$$

$$\left. m_{1,s} \leq i < m_1, m_{2,s} \leq k \leq m_2 \right\}.$$

(5.23)

$P\left(E^{RAMA,(s+1)}_{\{m_1,m_2\},M_{ac}}\right)$ *is given by averaging (5.22) over (5.6) as follows*

$$P\left(E^{RAMA,(s+1)}_{\{m_1,m_2\},M_{ac}}\right) = M_{ac}! \sum_{M_{ac}=0}^{\infty} \int_{z_1>\ldots>z_l>0}$$

$$P\left(E^{RAMA,(s+1)}_{\{m_1,m_2\},M_{ac}} \Big| G_{\{z_1,\cdots z_{M_{ac}}\}}\right) P\left(G_{\{z_1,\cdots z_{M_{ac}}\}}\right) dz_1\ldots z_{M_{ac}}.$$

(5.24)

We define the outage event of an active user, i.e., E^{RAMA}, to be event where none of its signal layers are successfully decoded by the BS. The outage performance of RAMA is shown in the following Theorem 5.2.

Theorem 5.2 *The exact expressions of the average outage probability and the average throughput of RAMA after sth SIC are given by (5.25) and (5.26), respectively.*

$$P\left(E^{RAMA}\right) = \frac{\sum_{M_{ac}=1}^{\infty} \left(P\left(K_{M_{ac}}\right) \sum_{m_1=1}^{M_{ac}} \left(\left(\sum_{m_2=1}^{M_{ac}} P\left(E^{RAMA,(s)}_{\{m_1,m_2\},M_{ac}}\right)\right) (M_{ac}-m_1+1) \right) \right)}{\sum_{M_{ac}=1}^{\infty} P\left(K_{M_{ac}}\right) M_{ac}},$$

(5.25)

$$T^{RAMA} = \sum_{M_{ac}=1}^{\infty} \left(P\left(K_{M_{ac}}\right) \sum_{m_1=1}^{M_{ac}} \left(\sum_{m_2=1}^{M_{ac}} \left(P\left(E^{RAMA,(s)}_{\{m_1,m_2\},M_{ac}}\right) (m_1 r_1^{RAMA} + m_2 r_2^{RAMA}) \right) \right) \right).$$

(5.26)

Proof The proof is similar to the proof of Theorem 5.1.

As aforementioned, RAMA introduces multiple layers at the transmitters, therefore the power and transmission rates allocation among different layers can act as an additional degree of freedom to match the statistical characteristics of interference, and to further enhance the network throughput of RAMA. The optimal power and transmission rates of RAMA is given by

$$\max_{\alpha, r_1^{\text{RAMA}}, r_2^{\text{RAMA}}} T^{\text{RAMA}},$$
$$\text{s.t. } r_1^{\text{RAMA}} + r_2^{\text{RAMA}} = r^{\text{conv}}. \tag{5.27}$$
$$0 \le \alpha \le 1.$$

Note that conv-GF only consider a single signal layer, and thus is not as adjustable as RAMA.

5.4.3   Comparisons

In this subsection, we compare the outage performance achieved by conv-GF and RAMA. First of all, we note that by setting $\alpha = 1$, $\varphi_1 = \phi$ and $\varphi_2 = 0$ in (5.27), the outage and throughput performance of RAMA is exactly the same as conv-GF. Therefore, it is straightforward that RAMA outperforms conv-GF with sophisticatedly designed parameters.

To visually illustrate the advantage of RAMA with respect to the outage performance, we compare the *complementary of the outage regions* of conv-GF and RAMA, i.e., $\left(\bigcup_{1 \le m \le M_{\text{ac}}} C_{m, M_{\text{ac}}}^{\text{conv}} \right)^c$ and $\left(\bigcup_{1 \le m_1, m_2 \le M_{\text{ac}}} C_{\{m_1, m_2\}, M_{\text{ac}}}^{\text{RAMA},(s)} \right)^c$, with $M_{\text{ac}} = 2$ and 3, as shown in Figs. 5.6 and 5.7, respectively. In the simulation, conv-GF and RAMA are set with the same transmission rates which takes value in $\{0.8, 1, 1.2\}$. The transmission rates and the power coefficients of RAMA is optimized according to (5.27). In Fig. 5.6, the black cycles represent all possible realizations of $\mathbf{H}$ with $M_{\text{ac}} = 2$, and the red crosses represent the realizations of $\mathbf{H}$ such that the signals of the two users are successfully recovered. We found that, with the increase of total transmission rate, i.e., ϕ, the outage performance of conv-GF becomes worse, while RAMA can achieve successful transmissions with almost all channel realizations. In Fig. 5.7, the black dots represent all possible realizations of $\mathbf{H}$ with $M_{\text{ac}} = 3$, and the red cycled points represents the realizations of $\mathbf{H}$ such that the three users can be successfully recovered. The advantage of RAMA over conv-GF with respect to the outage performance with $M_{\text{ac}} = 3$ is more significant than $M_{\text{ac}} = 2$, which validate the robustness of RAMA. We also observe that when the interference among users are severe, i.e., the areas selected by cycles in Figs. 5.6b, c, and 5.7b, c, conv-GF cannot ensure successful transmissions, while RAMA still achieve high throughput. To sum up, RAMA achieves high data rate in low interference region, while the robustness can also be assured in high interference region.

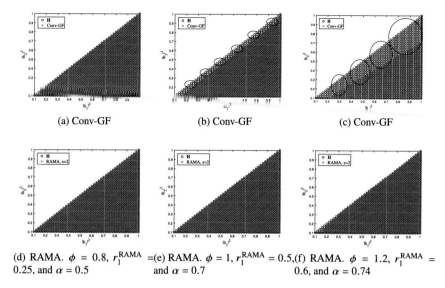

(a) Conv-GF   (b) Conv-GF   (c) Conv-GF

(d) RAMA. $\phi = 0.8$, r_1^{RAMA} =(e) RAMA. $\phi = 1$, $r_1^{RAMA} = 0.5$,(f) RAMA. $\phi = 1.2$, $r_1^{RAMA} =$
0.25, and $\alpha = 0.5$ and $\alpha = 0.7$ 0.6, and $\alpha = 0.74$

Fig. 5.6 Comparison on the complementary of the outage regions of conv-GF and RAMA with $M_{ac} = 2$. The SNR of each user is 10 dB

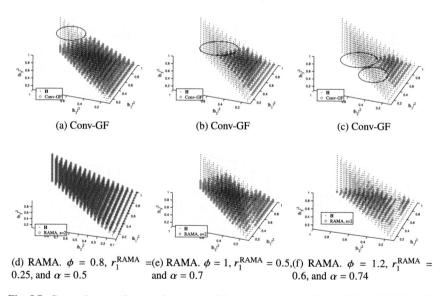

(a) Conv-GF   (b) Conv-GF   (c) Conv-GF

(d) RAMA. $\phi = 0.8$, r_1^{RAMA} =(e) RAMA. $\phi = 1$, $r_1^{RAMA} = 0.5$,(f) RAMA. $\phi = 1.2$, $r_1^{RAMA} =$
0.25, and $\alpha = 0.5$ and $\alpha = 0.7$ 0.6, and $\alpha = 0.74$

Fig. 5.7 Comparison on the complementary of the outage regions of conv-GF and RAMA with $M_{ac} = 3$. The SNR of each user is 10 dB

5.5 RAMA Amenable Constellations

In the above analysis, we have considered the ideal situation where Gaussian-distributed continuous alphabet is assumed at the transmitter. However, only finite alphabets can be deployed in practice. Therefore, we focus on the design and optimization of RAMA amenable constellations in this section.

The RAMA amenable constellations are composed of several sub-constellations, where each sub-constellation corresponds to a signal layer at transmitter. We call the equivalent channel experienced by each signal layer as a *sub-channel*, as mentioned in Sect. 5.3.2. To facilitate RAMA, our aim is to construct the sub-channels such that they have UEP. Corresponding to the two options in Step 3 of the RAMA scheme in Sect. 5.3.2, we propose two methods to design RAMA amenable constellations as well as the sub-channels in the following.

5.5.1 Overlapping Method

Corresponding to the option 1 in Step 3 of RAMA, we propose the overlapping method, where the composite constellation is constructed by overlapping several base constellations, i.e., $\{\mathcal{X}_{(\lambda_m^t, \vartheta_m^t)}\}^{n/\log_2\left(\left|\mathcal{X}_{(\lambda_m^t, \vartheta_m^t)}\right|\right)}$, as shown in (3). In Fig. 5.8, we show two examples of the RAMA amenable constellations, where BPSK and QPSK are employed as the basic building blocks. When the power coefficients, i.e., λ_1 and λ_2, of different base constellations are different, the bits in the composite constellation normally have different constellation constrained capacity. Therefore, we regard each bit (or several bits) as a sub-channel where one signal layer can be transmitted, as shown in Figs. 5.8a, b, respectively. We note that, with the proposed constellations,

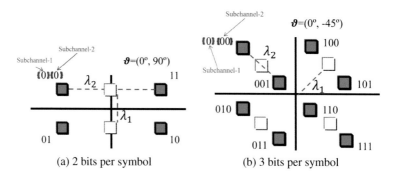

(a) 2 bits per symbol   (b) 3 bits per symbol

Fig. 5.8 Illustrations of RAMA amenable constellations, which are constructed by the overlapping method. λ_1 and λ_2 are the power coefficients for each base constellations, and ϑ is the rotation angle vector

even all signal layers are encoded with the same coding rate (to save the hardware resources), they still exhibit UEP property, which facilitates the SIC receiving of RAMA.

Furthermore, the power coefficients, i.e., λ_m^t, and rotation angles, i.e., ϑ_m^t shall be optimized to adapt to RAMA. In this chapter, we sequentially optimize these coefficients. The optimization of λ_m^t includes following three steps:

- Step 1: Define different reliability levels for different sub-channels, e.g., different BLER targets for the codewords transmitted in different sub-channels.
- Step 2: Map the reliability levels to the capacity of different sub-channels.
- Step 3: Adjust λ_m^t to meet the capacity requirements of different sub-channels.

With the fixed λ_m^t, ϑ_m^t should be optimized to achieve optimal constellation constraineds capacity. When the overlapped layers at transmitter are set to 2, optimal rotation angles can be derived by

$$\max_{\vartheta_m} m\left(\vartheta_m^t\right),$$ (5.28)

where $m\left(\vartheta_m\right)$ is given by (5.29) [22].

$$m\left(\vartheta_m\right) = \sum_{x_1 \in \mathcal{X}_m^t(\lambda_m, \vartheta_m^t)} \log\left(\sum_{x_2 \in \mathcal{X}_m^t(\lambda_m, \vartheta_m^t)} \frac{1}{\left|\mathcal{X}_m^t(\lambda_m^t, \vartheta_m^t)\right|^2} \exp\left(-\frac{1}{4\sigma^2}\|x_1 - x_2\|^2\right)\right)$$ (5.29)

5.5.2 Bundling Method

When high-order constellations are applied in RAMA, as illustrated in the option 2 of Step 3 in Sect. 5.3.2, we propose the bundling method to construct sub-channels, where different numbers of bits are bundled for different sub-channels. We show an example in Fig. 5.9, where a 16-QAM constellation is employed as the composite constellation of RAMA. We use the first bit as the sub-channel-1 and the rest three bits as the sub-channel-2. Suppose that high priority data stream and low priority data stream are transmitted in sub-channel-1 and sub-channel-2, respectively. To ensure that high priority data stream is better protected, low-rate channel coding should be adopted. After canceling the signal transmitted in sub-channel-1, the residual constellation is shown at the right-hand side of Fig. 5.9.

We note that the distances between the constellation points in the composite constellation cannot be arbitrarily adjusted. Therefore the UEP property are mainly offered by using different coding rates at different signal layers, which may lead to heavy burden on hardware resources. This fact makes the bundling method less flexible compared with the overlapping method.

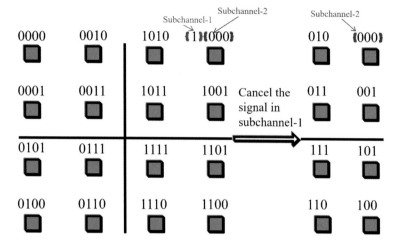

Fig. 5.9 An illustration of RAMA amenable constellation with 4 bits per symbol, which are constructed by the bundling method

5.6 Simulation Results

In this section, we evaluate the performance of the proposed RAMA scheme. First of all, we assume that Gaussian-distributed continuous alphabet and ideal SIC receiving are applied at the transmitter and the receiver, respectively. Based on these settings and the theoretical analysis in Sect. 5.4, we compare the outage and throughput performance of RAMA with that of the conv-GF [23]. Next, we validate the advantages of RAMA in practical situations, where the RAMA amenable constellations designed in Sect. 5.5 and realistic MMSE-SIC receiver are assumed.

5.6.1 Ideal Settings

As aforementioned, we assume Gaussian-distributed continuous alphabet and ideal SIC receiving. Figures 5.10 and 5.11 compare the analytical and simulation results of conv-GF and RAMA, with $M = 100$, $R_1 = 100$, and $R_2 = 10$. The average SNR of each user is assumed as 10 dB. The analytical results of conv-GF and RAMA, which are respectively shown by "□" and "○", is derived by (5.14) and (5.25), respectively, via Monte-Carlo sampling method. The upper bounds of the outage performance of conv-GF, which are shown by dashed lines, are derived by integrating the results in Appendix 3 into (5.14) and setting $P\left(E_{1,M_{ac}}^{conv}\right) = 1$, $M_{ac} > 3$. The upper bound is more tight when ϕ becomes larger, since the outage probability raises sharply when $M_{ac} > 3$. The simulation results show that RAMA can simultaneously achieve higher throughput and lower outage probability than conv-GF.

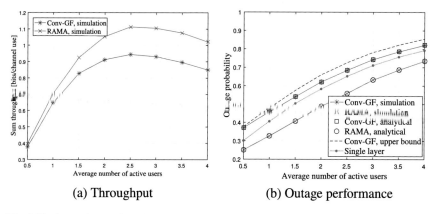

(a) Throughput   (b) Outage performance

Fig. 5.10 Comparisons of the throughput and outage performance between conv-GF and RAMA, with $\phi = 1.2$, $r_1^{\text{RAMA}} = 0.6$, $r_2^{\text{RAMA}} = 0.6$ and $\alpha = 0.74$

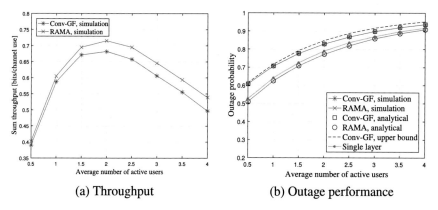

(a) Throughput   (b) Outage performance

Fig. 5.11 Comparisons of the throughput and outage performance between conv-GF and RAMA, with $\phi = 2$, $r_1^{\text{RAMA}} = 1$, $r_2^{\text{RAMA}} = 1$ and $\alpha = 0.75$

In the simulation, we consider a special case, namely "Single-layer", where only layer-1 of each user is transmitted in RAMA and layer-2 is assumed as noise. Figure 5.10b show that the outage performance can be enhanced compared with conv-GF, however, the gains are much smaller than RAMA. These results validate that the outage performance gain of RAMA comes from two aspects: first of all, the layer-1 of each user has high protection property due to its low coding rate and high power ratio; secondly, the layered structure facilitates the interference cancellation and further enhances the outage performance, since the users with small channel gains can benefit from the cancellation of highly protected layers of the users with large channel gains, and this is different from conv-GF where the signals of the users should be entirely recovered before cancellation.

5.6.2   Realistic Settings

In this subsection, we conduct a link level simulation of conv-GF and RAMA with realistic settings. We assume a single-cell OFDM-based uplink system with single antenna at both the BS and the user. The number of users is 100, and the activation probability varies from 0.002 to 0.06. The average received SNR of each user is assumed to take value from [4, 20] dB uniformly [25]. We also assume that the small scale channel coefficients follows Rayleigh fading, and the correlation coefficients among the small scale channels of the symbols within a transmission block is set as 0.2, 0.5, and 0.8, where the larger the correlation coefficient, the flatter the wireless channel.

For conv-GF, we apply 1/3 rate Turbo coding and QPSK modulation, while for RAMA, we assume $L_m = 2$ with 1/3 rate Turbo coding for both layers, and apply the constellation provided in Fig. 5.8a with $\lambda_1/\lambda_2 = 2, 4, 6$. The length of information bits is assumed as 1024. At the receiver, we apply Algorithm 5.2 to separate different users signals. Specifically, minimum mean square error (MMSE) detection is employed in the step 1 of Algorithm 5.2, which is given by [26]

$$\hat{x}_{m,l} = \left(h_m^t\right)^H \left(\sum_{I_j^t=1, j \neq m} h_m^t \left(h_m^t\right)^H + \sigma I^2 \right) y, \qquad (5.30)$$

where $\hat{x}_{m,l}$ is the estimated signal of mth user. Note that, other advanced demodulation technique, e.g. message passing algorithm (MPA), is not precluded in RAMA. The detailed simulation settings can be found in Table 5.1. Figures 5.12 and 5.13

Table 5.1 Simulation parameters

Parameters	Values or assumptions
User number	100
Activation probability	0.002:0.002:0.06
Waveform	OFDM
Average received SNR	Uniformly distributed in [4, 20] dB
Small scale channel model	Rayleigh fading, with correlation coefficients equals to 0.2, 0.5, 0.8
Transmission mode	TM 1 (SISO)
Length of information bits	1024
Channel coding	1/3 rate Turbo
Modulation of conv-GF	QPSK
Modulation of RAMA	Use Fig. 5.8a with $\lambda_1/\lambda_2 = 2, 4, 6$
Channel estimation	Ideal
Receiver	MMSE-SIC

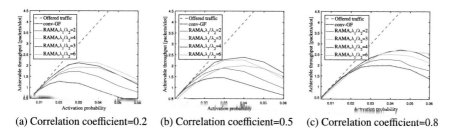

(a) Correlation coefficient=0.2   (b) Correlation coefficient=0.5   (c) Correlation coefficient=0.8

Fig. 5.12 Comparison on the throughput performance between conv-GF and RAMA

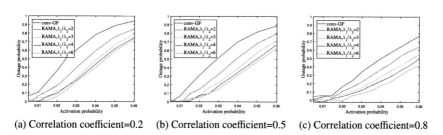

(a) Correlation coefficient=0.2   (b) Correlation coefficient=0.5   (c) Correlation coefficient=0.8

Fig. 5.13 Comparison on the outage probability between conv-GF and RAMA

compare the throughput and the outage performance of conv-GF and RAMA. Generally, the system throughput enhances with the increase of the channel correlation coefficients, i.e. γ. With elaborately selected power coefficients of constellation, RAMA achieves larger throughput, as well as lower outage probability than conv-GF. This result also reflects that the fairness among grant-free users is improved with RAMA. Moreover, the performance gains of RAMA are more significant when the underlaying physical channel is not flat. Besides, we also observe that RAMA achieves high robustness when the activation probability goes larger (e.g. $\gamma = 0.06$), while conv-GF experiences significant drop in total throughput.

5.7 Conclusions

In this chapter, we have proposed the RAMA scheme for uplink grant-free data transmission. By employing layered signal structure at the transmitter and intra- and inter-user SIC at the receiver, the proposed RAMA scheme achieves significant throughput and outage performance gain over conv-GF, which have been validated by analysis and simulations. The actual transmission data rate can adapt to the actual channel conditions of active users, which cannot be foreseen. RAMA also achieves high robustness when the activation probability of the users is large. Despite all this improvement, we discuss some open issues of RAMA that are worth further study in the following.

Joint design with spreading-based NOMA: In this chapter, we have mainly focused on the grant-free transmission based on the power domain NOMA, where symbol level spreading is not included. RAMA can also co-deploy with other state-of-art NOMA schemes, where the main idea is to incorporate multiple independent signal layers at the transmitter.

Location-based access: Although the grant-free user cannot acquire the accurate channel information, they may estimate their large scale channel coefficients, e.g. via reference signal receiver power (RSRP) at the downlink. This side information may serve as an important factor based on which the grant-free users choose suitable power and transmission data rates for different layers in RAMA.

Error propagation: Since multiple layers are introduced in RAMA, a natural problem is the error propagation among different layers during SIC receiving. This issue may be addressed by deploying the joint-detection based receiver, which also requires further study.

Appendix 1 Proof of Lemma 5.1

As assumed, the users are uniformly distributed in the cell. Therefore the PDF of the distance between the users and the BS is given by

$$f_r(v) = \frac{2v}{R_1^2 - R_2^2}, \; R_2 \le v \le R_1. \tag{5.31}$$

The PDF of the square of the distance, i.e., $x = v^2$, is

$$
\begin{aligned}
f_{r^2}(y) &= f_r\left(y^{-\frac{1}{2}}\right)\left(y^{-\frac{1}{2}}\right)'_y \\
&= \frac{2x^{\frac{1}{2}}}{R_1^2 - R_2^2}\left(\frac{1}{2}x^{-\frac{1}{2}}\right) = \frac{1}{R_1^2 - R_2^2}, \; R_2^2 \le y \le R_1^2.
\end{aligned}
\tag{5.32}
$$

The square of the magnitude of the small scale fading coefficient follows the $\chi^2(v)$ distribution with two degrees of freedom, i.e. $v = 2$, and is given by

$$f_{|g|^2}(x) = \frac{1}{2}e^{-\frac{x}{2}}. \tag{5.33}$$

Then by integrating (5.32) and (5.33) in (5.2), we have

$$f_{|h|^2}(z) = \int_0^{\infty} f_{|g|^2}(zy) f_{r^2}(y) y \, dy$$

$$= \int_{d_2^2}^{d_1^2} \frac{1}{2} e^{-\frac{zy}{2}} \frac{1}{R_1^2 - R_2^2} y \, dy$$ (5.34)

$$= \frac{2}{z^2 \left(R_1^2 - R_2^2\right)} \int_{d_2^2}^{d_1^2} e^{-\frac{zy}{2}} \left(\frac{zy}{2}\right) d\left(\frac{zy}{2}\right)$$

$$= \frac{1}{z^2 \left(R_1^2 - R_2^2\right)} \left(e^{-\frac{R_2^2 z}{2}} \left(R_2^2 z + 2\right) - e^{-\frac{R_1^2 z}{2}} \left(R_1^2 z + 2\right)\right),$$

where the last equation is due to $\int e^{-t} t \, dt = -t e^{-t} - e^{-t} + C$.

The CDF of $f_{|h|^2}(z)$ is calculated as follows. We note that the indefinite integration of $f_{|h|^2}(z)$ is given by

$$h(z) = \int f_{|h|^2}(z) dz = \frac{2e^{-(R_1^2 z)/2} - 2e^{-(R_2^2 z)/2}}{z\left(R_1^2 - R_2^2\right)},$$ (5.35)

where $h(z) \to -1$, $z \to 0^+$. The CDF of $|h|^2$ is given as

$$F_{|h|^2}(z) = h(z) - h(0) = 1 + h(z).$$ (5.36)

$$P\left(E_{m,M_{ac}}^{conv,t}\right) = M_{ac}! \times \int_0^{+\infty} f_{|h'|^2}(z_{M_{ac}}) \int_{z_{M_{ac}-1}=z_{M_{ac}}}^{+\infty} f_{|h'|^2}(z_{M_{ac}-1}) \cdots$$

$$\times \int_{z_{m+1}=z_{m+2}}^{+\infty} f_{|h'|^2}(z_{m+1}) \times \int_{z_m=z_{m+1}}^{\phi(\sum_{j=m+1}^{M_{ac}} z_j + \sigma^2/P)} f_{|h'|^2}(z_m) \times \int_{z_{m-1}=\max\left(z_m, \phi(\sum_{j=m}^{M_{ac}} z_j + \sigma^2/P)\right)}^{+\infty} f_{|h'|^2}(z_{m-1}) \cdots$$

$$\times \left[\underbrace{\int_{z_3=\max\left(z_4, \phi(\sum_{j=4}^{M_{ac}} z_j + \sigma^2/P)\right)}^{+\infty} f_{|h'|^2}(z_3) \int_{z_2=\max(z_3, \frac{\phi}{1-\phi}(\sum_{j=3}^{M_{ac}} + \sigma^2/P))}^{+\infty} f_{|h'|^2}(z_2) \int_{z_1=z_2}^{+\infty} f_{|h'|^2}(z_1)}_{I_1}\right.$$

$$\left. + \underbrace{\int_{z_3=\max\left(z_4, \phi(\sum_{j=4}^{M_{ac}} z_j + \sigma^2/P)\right)}^{+\infty} f_{|h'|^2}(z_3) \int_{z_2=\max(z_3, \phi(\sum_{j=3}^{M_{ac}} + \sigma^2/P))}^{\frac{\phi}{1-\phi}(\sum_{j=3}^{M_{ac}} + \sigma^2/P)} f_{|h'|^2}(z_2) \int_{z_1=\phi(\sum_{j=2}^{M_{ac}} z_j + \sigma^2/P)}^{+\infty} f_{|h'|^2}(z_1)}_{I_2} \right] dz_1 \cdots z_{M_{ac}}$$

(5.37)

$$I_1 = \int\limits_{z_3=\max\left(z_4,\phi\left(\sum_{j=4}^{M_{ac}} z_j+\sigma^2/P\right)\right)}^{+\infty} f_{|h^t|^2}(z_3) \int\limits_{z_2=\frac{\phi}{1-\phi}\left(\sum_{j=3}^{M_{ac}}+\sigma^2/P\right)}^{+\infty} f_{|h^t|^2}(z_2) \int\limits_{z_1=z_2}^{+\infty} f_{|h^t|^2}(z_1).$$

$$(5.38)$$

$$I_2 = \int\limits_{z_3=z_4}^{+\infty} f_{|h^t|^2}(z_3) \int\limits_{z_2=z_3}^{\frac{\phi}{1-\phi}\left(\sum_{j=3}^{M_{ac}}+\sigma^2/P\right)} f_{|h^t|^2}(z_2) \int\limits_{z_1=\phi\left(\sum_{j=2}^{M_{ac}} z_j+\sigma^2/P\right)}^{+\infty} f_{|h^t|^2}(z_1)$$

$$(5.39)$$

$$+ \int\limits_{z_3=\max\left(z_4,\phi\left(\sum_{j=4}^{M_{ac}} z_j+\sigma^2/P\right)\right)}^{+\infty} f_{|h^t|^2}(z_3) \int\limits_{z_2=\phi\left(\sum_{j=3}^{M_{ac}}+\sigma^2/P\right)}^{\frac{\phi}{1-\phi}\left(\sum_{j=3}^{M_{ac}}+\sigma^2/P\right)} f_{|h^t|^2}(z_2) \int\limits_{z_1=\phi\left(\sum_{j=2}^{M_{ac}} z_j+\sigma^2/P\right)}^{+\infty} f_{|h^t|^2}(z_1).$$

Appendix 2 Recursive Expression of (5.12)

We assume that $\phi \geq 1/2$ in the following derivation, where the derivation with $\phi < 1/2$ follows the same approach.

To eliminate the max operations and derive the exact expressions of (5.12), we consider two cases, i.e., $z_2 \geq \phi\left(\sum_{j=2}^{M_{ac}}+\sigma^2/P\right)$ and $z_2 < \phi\left(\sum_{j=2}^{M_{ac}}+\sigma^2/P\right)$, where $z_2 \geq \frac{\phi}{1-\phi}\left(\sum_{j=3}^{M_{ac}}+\sigma^2/P\right)$, and $z_2 \geq \frac{\phi}{1-\phi}\left(\sum_{j=3}^{M_{ac}}+\sigma^2/P\right)$, respectively. Therefore, (5.12) is given by (5.37), where I_1 and I_2 can be further simplified in the following.

Observe that $1 > \phi \geq 1/2$, I_1 can be simplified as derived in (5.38). Besides, to eliminate the max operation in I_2, we consider two cases, i.e., $z_3 \geq \phi\left(\sum_{j=3}^{M_{ac}}+\sigma^2/P\right)$ and $z_3 < \phi\left(\sum_{j=3}^{M_{ac}}+\sigma^2/P\right)$, where $z_3 \geq \frac{\phi}{1-\phi}\left(\sum_{j=4}^{M_{ac}}+\sigma^2/P\right)$, and $z_3 \geq \frac{\phi}{1-\phi}\left(\sum_{j=4}^{M_{ac}}+\sigma^2/P\right)$, respectively. And I_2 is given by (5.39). Till now, the max operations related to z_1 and z_2 are completely eliminated. By recursively conducting the procedures between (5.37) and (5.39), the exact expression of (5.12) can be derived.

Appendix 3 Proof of Proposition

5.3

Assuming $M_{ac} = 1$, $P\left(E_{1,1}^{conv}\right)$ is given by

$$P\left(E_{1,1}^{\text{conv}}\right) = \int_{z_1=0}^{\phi\sigma^2/P} f_{|h'|^2}(z_1)dz_1 = F_{|h'|^2}(\phi\sigma^2/P)$$

$$= 1 + \frac{2e^{-(R_1^2\phi\sigma^2/P)/2} - 2e^{-(R_2^2\phi\sigma^2/P)/2}}{\phi\sigma^2\left(R_1^2 - R_2^2\right)/P}.$$

(5.40)

Furthermore, if $M_{\text{ac}} = 2$, $P\left(E_{1,2}^{\text{conv}}\right)$ and $P\left(E_{2,2}^{\text{conv}}\right)$ are given by

$$P\left(E_{1,2}^{\text{conv}}\right) = 2! \int_{z_2=0}^{+\infty} f_{|h'|^2}(z_2) \int_{z_1=z_2}^{\phi(z_2+\sigma^2/P)} f_{|h'|^2}(z_2)dz_1dz_2$$

$$= 2 \int_{z_2=0}^{+\infty} f_{|h'|^2}(z_2) \left(F_{|h'|^2}(\phi(z_2 + \sigma^2/P)) - F_{|h'|^2}(z_2)\right) dz_2$$

$$= 2\left(\mathcal{F}(+\infty) - \mathcal{F}(0)\right) - 1.$$

(5.41)

and

$$P\left(E_{2,2}^{\text{conv}}\right) =$$

$$2! \int_{z_2=0}^{\phi\sigma^2/P} f_{|h'|^2}(z_2) \int_{z_1=\phi(z_2+\sigma^2/P)}^{+\infty} f_{|h'|^2}(z_1)dz_1dz_2$$

$$= 2 \int_{z_2=0}^{\phi\sigma^2/P} f_{|h'|^2}(z_2) \left(F_{|h'|^2}(+\infty) - F_{|h'|^2}(\phi(z_2 + \sigma^2/P))\right) dz_2$$

$$= 2\left(F_{|h'|^2}(\phi(\sigma^2/P)) - \mathcal{F}(\phi(\sigma^2/P)) + \mathcal{F}(0)\right).$$

(5.42)

respectively, where

$$\mathcal{F}(z) = \int f_{|h'|^2}(z) F_{|h'|^2}(\phi(z + \sigma^2/P))dz.$$

(5.43)

Assume high SNR and $\phi = 1$, (5.43) can be calculated as $\mathcal{F}(z) = \frac{1}{2}(F(z))^2$.

When $M_{\text{ac}} = 3$, $P\left(E_{1,3}^{\text{conv}}\right)$, $P\left(E_{2,3}^{\text{conv}}\right)$ and $P\left(E_{3,3}^{\text{conv}}\right)$ can be derived with the similar approaches, which is omitted due to the space limitation.

References

1. TR 38.913, Study on scenarios and requirements for next generation access technologies. 3GPP, 2015. (Release 14)

2. K. Au, L. Zhang, H. Nikopour, et al., Uplink contention based SCMA for 5G radio access, in *Globecom Workshops (GC Workshops)* (IEEE, Austin TX, USA, 2014), pp. 900–905
3. C. Bockelmann, N. Pratas, H. Nikopour et al., Massive machine-type communications in 5G: physical and MAC-layer solutions. IEEE Commun. Mag. **54**(9), 59–65 (2016)
4. Y. Yuan, Z. Yuan, G. Yu et al., Non-orthogonal transmission technology in LTE evolution. IEEE Commun. Mag. **54**(7), 68–74 (2016)
5. Y.J. Choi, K.G. Shin, Joint collision resolution and transmit-power adjustment for Aloha-type random access. Wirel. Commun. Mob. Comput. **13**(2), 184–197 (2013)
6. C. Zhu, L. Shu, T. Hara et al., A survey on communication and data management issues in mobile sensor networks. Wirel. Commun. Mob. Comput. **14**(1), 19–36 (2014)
7. E. Paolini, G. Liva, M. Chiani, Coded slotted ALOHA: A graph-based method for uncoordinated multiple access. IEEE Trans. Inform. Theory. **61**(12), 6815–6832 (2015)
8. A. Wyner, Recent results in the Shannon theory. IEEE Trans. Inform. Theory **20**(1), 2–10 (1974)
9. M. Médard, J. Huang, A.J. Goldsmith et al., Capacity of time-slotted ALOHA packetized multiple-access systems over the AWGN channel. IEEE Trans. Wirel. Commun. **3**(2), 486–499 (2004)
10. P. Minero, M. Franceschetti, N.C. David, Random access: An information-theoretic perspective. IEEE Trans. Inform. Theor. **58**(2), 909–930 (2012)
11. R.H. Etkin, N.C. David, H. Wang, Gaussian interference channel capacity to within one bit. IEEE Trans. Inform. Theory **54**(12), 5534–5562 (2008)
12. ZTE, R1-1701608, Performance evaluation of nonorthogonal multiple access in 2-step random access procedure. 3GPP (2017)
13. Y. Wu, J. Fang, Large-scale antenna-assisted grant-free non-orthogonal multiple access via compressed sensing (2016). arXiv preprint arXiv:1609.00452
14. N. Zhang, J. Wang, G. Kang et al., Uplink nonorthogonal multiple access in 5G systems. IEEE Commun. Lett. **20**(3), 458–461 (2016)
15. H. Zheng, X. Li, N. Ye, Random subchannel selection of store-carry and forward transmissions in traffic hotspots. IEEE Commun. Lett. **21**(9), 2073–2076 (2017)
16. A. El Gamal, Y.H. Kim, *Network Information Theory* (Cambridge University Press, 2011), pp. 151–158
17. B. Rimoldi, R. Urbanke, A rate-splitting approach to the Gaussian multiple-access channel. IEEE Trans. Inform. Theory **42**(2), 364–375 (1996)
18. H. Joudeh, B. Clerckx, Rate-splitting for max-min fair multigroup multicast beamforming in overloaded systems. IEEE Trans. Wirel. Commun. **16**(11), 7276–7289 (2017)
19. B. Clerckx, H. Joudeh, C. Hao et al., Rate splitting for MIMO wireless networks: a promising PHY-layer strategy for LTE evolution. IEEE Commun. Mag. **54**(5), 98–105 (2016)
20. NTT DOCOMO, INC., R1-1713952, UL data transmission without UL grant. 3GPP (2017)
21. Intel, R1-1712592:, UL data transmission without grant. 3GPP (2017)
22. N. Ye, A. Wang, X. Li, W. Liu, X. Hou, H. Yu, On constellation rotation of NOMA with SIC receive. IEEE Commun. Lett. **22**(3), 514–517 (2018)
23. NTT DOCOMO, INC., R1-167392, Discussion on multiple access for UL mMTC. 3GPP (2016)
24. A. Benjebbour, Y. Saito, Y. Kishiyama, et al., Concept and practical considerations of non-orthogonal multiple access (NOMA) for future radio access, in *2013 International Symposium on Intelligent Signal Processing and Communications Systems (ISPACS)* (IEEE, Okinawa, Japan, 2013), pp. 770–774
25. Z. Yuan, G. Yu, W. Li, et al., Multi-user shared access for internet of things, in *IEEE Vehicular Technology Conference* (IEEE, 2016), pp. 1–5
26. K. Higuchi, A. Benjebbour, Non-orthogonal multiple access (NOMA) with successive interference cancellation for future radio access. IEICE Trans. Commun. **98**(3), 403–414 (2015)

Chapter 6
Artificial Intelligence-Enhanced Multiple Access

Abstract In this chapter, we discuss the artificial intelligence-enhanced multiple access. Section 6.1 introduces the motivation of applying deep multi-task learning to end-to-end optimization of NOMA. Section 6.2 describes the system model and formulate the end-to-end optimization problem. Section 6.3 proposes the general DeepNOMA framework. Sections 6.4 and 6.5 propose DeepMAS and DeepMUD. Section 6.6 presents the experiment and simulation results. Section 6.7 illustrates the conclusions.

6.1 Introduction

Non-orthogonal multiple access (NOMA) has been regarded as a key enabling technology to fulfill the massive connectivity requirement in future Internet of Things (IoT) [1, 2]. With the elaborately designed multiple access signatures (MASs), NOMA can multiplex various signal streams with controllable mutual interference. Meanwhile, advanced multi-user detectors (MUDs) are applied to distinguish the superimposed signal streams. While NOMA has succeeded in providing massive connectivity and high spectral efficiency, it is challenging to jointly optimize MAS and MUD, due to the highly-condensed signal structure and the intractable system model. Hence, it is pivotal to develop efficient methods for further performance enhancement of NOMA.

In the most recent decade, deep learning (DL) has achieved great success in solving very complicated optimization problems in a data-driven fashion [3]. The booming of DL has also shed new light on the end-to-end optimization of wireless communication technologies in complicated scenarios [4–9]. Deep auto-encoding (DAE) structure is employed to model the end-to-end communication system, while deep neural network (DNN) is used as a universal function approximator to mimic the target mappings and to extract the efficient hidden features of signals. Nevertheless, existing DL is normally designed for human-level tasks, such as neural language processing and computer vision [3], and may not ideally suit the problems in wireless communications, especially in NOMA cases. Our work aims to bridge this gap by developing a unified DL framework of NOMA for further performance enhancements.

6.1.1 Related Work and Motivation

The concept of NOMA originates from multi-user information theory, where super-position transmission and successive interference cancellation (SIC) detection have been developed to achieve the outer bounds of multiple access channel (MAC) capacity region [10]. However, the deployment of NOMA is not made practical until the recent advent of powerful chips, thanks to Moore's Law, which are capable of decoding superimposed signals with iterative algorithms [11]. Nowadays, NOMA has been extensively researched towards massive connectivity in IoT [12]. The key design aspects of NOMA are related to MAS and MUD, where MAS determines the performance upper bound and MUD decides how close that bound can be approached.

MASs have been designed in various domains, such as power and code domains, to control the inter-user interference in NOMA. In power-domain NOMA, the power-level is recognized as user-specific signature; while in code-domain NOMA, the bit-to-symbol sequence mapping is of great importance. Existing code-domain NOMA variants can be categorized according to sequence characteristics, including the sparsity, e.g., dense [13] and sparse [14, 15], the linearity, e.g., linear [13] and non-linear [14], as well as the property of sequence, e.g., Welch-bound-equality (WBE) [16] and equiangular tight frames [17]. As efficient approaches to design good MAS, communication signal processing technology and information theory have been extensively investigated to maximize the minimum Euclidean distance [18] and maximize the constellation-constrained capacity [19], respectively. Dedicated techniques such as rotation [19, 20], projection [21], permutation [14], and interleaving [18] are developed to achieve the above goals. Nonetheless, these designs require many human-crafted works and are usually restricted in certain scenarios.

Great research efforts have also been made with regard to advanced MUD design [22], where graph model-based methods [23, 24], such as message passing algorithm (MPA), and interference cancellation (IC)-based methods, such as SIC [25], act as core driving forces. The concept of IC has played a vital role in various variants of MUDs to mitigate the inter-user interference and enhance the detection accuracy [23–27].

As aforementioned, existing research routes about NOMA normally isolate the design of the transceivers. This divide-and-conquer design philosophy may not satisfy the diversified, complicated and heterogeneous demands on multiple access in the future IoT [11]. A unified framework of NOMA is needed to simultaneously design good transceivers. Nonetheless, it is still mathematically intractable to conduct end-to-end optimization of NOMA, since multiple correlated signals make this problem hard to model and solve [8].

Fortunately, the recent booming of DL technology has shed a new light in solving very complicated, even intractable problem in a data-driven fashion [28]. While DL has proven its great value in channel encoding and decoding [29, 30], multiple input multiple output signal detection [31–33], and end-to-end communication system design [34, 35], only a few researches have studied the deployment of DL in enhancing NOMA. Authors in [36] propose a joint detection and decoding method of NOMA by unfolding MPA and belief-propagation decoding into DNN. In [37], the authors model a sparse code multiple access (SCMA) system as a DAE in additive white Gaussian noise channel, and propose codebooks by minimizing the signal reconstruction loss without any human-crafted works. Meanwhile, a recurrent DNN-based NOMA is proposed in [38] to address the varying channel conditions. Our previous work proposes a DL-aided grant free NOMA which takes the advantage of the random user activation phenomenon [28]. Besides, DL-based resource allocation of NOMA has also been studied [39].

Despite the above works, there still lacks a unified DL framework for further physical layer enhancement of NOMA. To this end, our paper aims to exploit the joint benefits of DL technology and communication-domain expertises in the transceiver optimization of NOMA.

6.1.2 Contributions

In this chapter, we propose a unified DL framework for end-to-end optimization of NOMA. We regard the overlapped transmissions as multiple learning tasks and recast the NOMA system with the multi-task DNN. In our framework, DNN acts as an underlaying universal function approximator providing strong learning ability, while sophisticated DNN structures are then designed based on multiple-access communication models for better learning efficiency. This framework provides new insights into a principled route to improving NOMA with DL.

The detailed contributions are summarized as follows:

1. We propose a multi-task DNN framework of NOMA, namely DeepNOMA, by treating non-orthogonal transmissions as multiple distinctive but correlated tasks. DeepNOMA uses an auto-encoding structure, which consists of a channel module, a signature mapping module, namely DeepMAS, and a detection module, namely DeepMUD. Data-driven end-to-end optimization on the transceivers is enabled by minimizing the overall cross-entropy reconstruction loss. A multi-task balancing technique is also developed to guarantee fairness among users and to avoid local optima.
2. We propose a model-based DeepMAS to reduce the implementation complexity. We embed the desired geometrical shapes of the transmit symbols into DNNs, such that the proposed DeepMAS only generates signatures with regular shapes. We also propose a parameter initialization principle which makes the statistical distributions of the composite signal to approach Gaussian under Kullback-Leibler (KL) divergence measure.
3. We design a novel network structure for DeepMUD, inspired from multi-user information theory, to exploit the superimposed signal structure of NOMA. We firstly propose an IC-enabled DNN (ICNN) which is shown to have enhanced representational power while dealing with superimposed signals. Then we propose a design of DeepMUD by incorporating ICNN as inter-task connections.

DeepMUD achieves a good tradeoff between the universality of DNN and the specialty of IC structure, which makes it a universal low-complexity framework for NOMA signal detections.

4. We implement DeepNOMA, evaluate its learning performance, and compare it with state-of-the-art NOMA schemes under various channel models. Link-level simulation results show the benefits of DeepNOMA in both transmission accuracy and computational complexity.[1]

6.2 System Model and Problem Formulation

6.2.1 Uplink NOMA System Model

Consider a synchronized uplink NOMA system with N users and K orthogonal resource elements (REs). The system overloading factor is defined as N/K, where $N/K > 100\%$ is normally considered in NOMA to achieve higher spectrum efficiency than orthogonal multiple access (OMA). The n-th user, $n = 1 \cdots N$, wishes to communicate a $\log_2(M)$-bit message $\mathcal{M}_n$ to the receiver, where $\mathcal{M}_n \in \{1, \cdots, M\}$ is the index of the message following the distribution of $P(\mathcal{M}_n)$, i.e., $\mathcal{M}_n \sim P(\mathcal{M}_n)$. For the n-th user, $\mathcal{M}_n$ is then mapped to a K-dimensional complex symbol $\mathbf{x}_n = [x_{n,k}]_{k=1}^{K} \in \mathbb{C}^K$ according to the the user-specific MAS mapping function f_n, which is defined as

$$f_n : \mathcal{M}_n \rightarrow \mathbf{x}_n \in \mathcal{X}_n \subset \mathbb{C}^K, \tag{6.1}$$

where $\mathcal{X}_n = \{\mathcal{X}_{n,1}, \cdots, \mathcal{X}_{n,M}\}$ is defined as the alphabet of transmission symbols with $|\mathcal{X}_n| = M$. In this chapter, we normalize the transmit power of each symbol as $\|\mathbf{x}_n\| = 1$. Define $\mathbf{h}_n = [h_{n,k}]_{k=1}^{K} \in \mathbb{C}^K$ as the channel coefficient vector of the n-th user over K REs. The received signal at the base station (BS) is given by

$$\mathbf{y} = \sum_{n=1}^{N} \sqrt{P_n}\,\mathrm{diag}\,(\mathbf{h}_n)\,\mathbf{x}_n + \mathbf{n}, \tag{6.2}$$

where P_n is the transmit power of the n-th user and $\mathbf{n} \sim \mathcal{CN}(0, \sigma_0^2\mathbf{I})$ denotes the additive white Gaussian noise (AWGN) vector with variance σ_0^2.

At the receiver, the distinctions among MASs are exploited by MUD to retrieve source messages. We regard the N superimposed transmissions as N distinctive but

---

[1] Notations: Normal lower-case, bold lower-case and bold upper-case symbols denote the scalars, vectors and matrices, respectively. $\mathbb{R}$ and $\mathbb{C}$ denote the fields of real and complex number, respectively. $[x_k]_{k=1}^{K}$ denotes the K-dimensional column vector where x_k is the k-th element. $\mathrm{diag}(\mathbf{x})$ represents the diagonal matrix with the diagonal specified by vector $\mathbf{x}$. $|\cdot|$ denotes the modulus of a scalar or the cardinality of a set, and $\|\cdot\|$ represents the ℓ_2-norm. $\mathcal{CN}(\boldsymbol{\mu}, \boldsymbol{\Sigma})$ represents circular symmetric complex Gaussian distribution with mean $\boldsymbol{\mu}$ and covariance $\boldsymbol{\Sigma}$.

correlated tasks. Thus, the objective of the n-th task is to obtain an accurate estimation of $\mathcal{M}_n$, as defined in

$$g_n : \mathbf{y} \to \hat{\mathcal{M}}_n \in \{1, \cdots, M\}, \tag{6.3}$$

where g_n represents the receiver which maps the received signal $\mathbf{y}$ to the estimated source message $\hat{\mathcal{M}}_n$. The average message error rate (MER) of the n-th task is then defined as

$$P_e^{(n)} = \mathbb{E}_{P(\mathcal{M}_n)} \left[\Pr\{\mathcal{M}_n \neq \hat{\mathcal{M}}_n\} \right]. \tag{6.4}$$

6.2.2 Problem Formulation

Due to the limited radio resources, multiple tasks in NOMA conflict with each other, which requires an end-to-end joint optimization. We adopt the Bayesian maximum a posteriori (MAP) perspective and formulate a variational optimization problem for joint transceiver optimization.

Assume that we have found the collection of the optimal MAS mapping functions $[f_i^*]_{i=1}^N$, we now focus on the design of MUD. For the n-th transmission task, we expect to find an optimal receiver g_n^* which maximizes the posterior probability mass function (PMF) of the transmitted message. Denote $\mathbf{y}_t$ as the received signal in the t-th time-slot, $t = 1, 2, \cdots$, and define $P(\mathcal{M}_{n,t}|\mathbf{y}_t; g_n)$ as posterior probability that g_n successfully recovers the correct source message $\mathcal{M}_{n,t}$ given $\mathbf{y}_t$. Given $[f_i^*]_{i=1}^N$, the MUD design problem of the n-th task is then formulated by averaging over all transmission time-slots, as follows

$$
\begin{aligned}
g_n^* &= \underset{g_n}{argmax} \prod_t P(\mathcal{M}_{n,t}|\mathbf{y}_t; g_n) \\
&= \underset{g_n}{argmax} \log \left(\prod_t P(\mathcal{M}_{n,t}|\mathbf{y}_t; g_n) \right) \\
&= \underset{g_n}{argmax} \sum_t \log P(\mathcal{M}_{n,t}|\mathbf{y}_t; g_n) \\
&\overset{(a)}{=} \underset{g_n}{argmax} \, \mathbb{E}_{P\left(\mathcal{M}_n, \mathbf{y}; [f_i^*]_{i=1}^N\right)} \left[\log P(\mathcal{M}_n|\mathbf{y}; g_n) \right],
\end{aligned}
\tag{6.5}
$$

where g_n is the variational function to be optimized. Note that, given $[f_i^*]_{i=1}^N$, each pair of $(\mathcal{M}_{n,t}, \mathbf{y}_t)$ is independent and identically distributed (i.i.d.) for each t. Therefore, (a) holds since averaging over $(\mathcal{M}_{n,t}, \mathbf{y}_t)$, $\forall t$, is equivalent to taking expectation with respect to the joint distribution of $(\mathcal{M}_n, \mathbf{y})$, i.e., $P(\mathcal{M}_n, \mathbf{y}; [f_i^*]_{i=1}^N)$. However, since f_i^* in (6.5) is actually unknown, we replace f_i^* with f_i, which is also treated as the variational function to be optimized.

We now write the objective function of (6.5) as $\mathcal{L}_n\left([f_i]_{i=1}^N, g_n\right)$ and simplify it. Define $\mathcal{M} = [\mathcal{M}_i]_{i=1}^N$ as the collection of the messages transmitted by N users, and $\mathcal{M}_n^C$ as the messages transmitted by others except the n-th user. $\mathcal{L}_n\left([f_i]_{i=1}^N, g_n\right)$ is simplified as follows

$$
\begin{aligned}
\mathcal{L}_n\left([f_i]_{i=1}^N, g_n\right) &= \int \log P(\mathcal{M}_n|\mathbf{y}; g_n) P\left(\mathcal{M}_n, \mathbf{y}; [f_i]_{i=1}^N\right) d\mathcal{M}_n d\mathbf{y} \\
&\overset{(a)}{=} \int \log P(\mathcal{M}_n|\mathbf{y}; g_n) P(\mathcal{M}_n) P(\mathbf{y}|\mathcal{M}_n; [f_i]_{i=1}^N) d\mathcal{M}_n d\mathbf{y} \\
&\overset{(b)}{=} \int \log P(\mathcal{M}_n|\mathbf{y}; g_n) P(\mathcal{M}_n) \\
&\quad \times \left(\int P(\mathbf{y}|\mathcal{M}_n, \mathcal{M}_n^C; [f_i]_{i=1}^N) P(\mathcal{M}_n^C) d\mathcal{M}_n^C \right) d\mathcal{M}_n d\mathbf{y} \\
&= \int P(\mathcal{M}) \left(\int \log P(\mathcal{M}_n|\mathbf{y}; g_n) P(\mathbf{y}|\mathcal{M}; [f_i]_{i=1}^N) d\mathbf{y} \right) d(\mathcal{M}) \\
&= \mathbb{E}_{P(\mathcal{M})} \mathbb{E}_{P(\mathbf{y}|\mathcal{M};[f_i]_{i=1}^N)} \left[\log P(\mathcal{M}_n|\mathbf{y}; g_n) \right],
\end{aligned}
\tag{6.6}
$$

where $P(\mathcal{A}|\mathcal{B})$ represents the conditional probability distribution of $\mathcal{A}$ given $\mathcal{B}$, and $P(\mathcal{M})$ is the joint distribution of $\mathcal{M}$ given by $P(\mathcal{M}) = \prod_{i=1}^N P(\mathcal{M}_i) = P(\mathcal{M}_n) P(\mathcal{M}_n^C)$. Here, step-(a) is due to the conditional probability formula, and step-(b) is due to the law of total probability.

Finally, we consider all N transmission tasks and present the problem formulation of the joint NOMA transceiver optimization as follows

$$
\mathcal{P}1 : \left[f_i^*\right]_{i=1}^N, \left[g_i^*\right]_{i=1}^N = \underset{f_n, g_n, n=1 \cdots N}{argmax} \mathcal{L}\left([f_i]_{i=1}^N, [g_i]_{i=1}^N\right),
\tag{6.7}
$$

$$
\text{s.t. (6.1), (7.1), and (6.3),}
$$

where

$$
\mathcal{L}\left([f_i]_{i=1}^N, [g_i]_{i=1}^N\right) = \sum_{n=1}^N w_n \mathcal{L}_n\left([f_i]_{i=1}^N, g_n\right),
\tag{6.8}
$$

is the objective function derived by summing over the optimization targets of the N tasks in (6.6) with the weight coefficient w_n.

Although solving $\mathcal{P}1$ provides the optimal transceivers under the perspective of MAP estimation, it is actually non-trivial to solve this problem. The major difficulties lay in the complicated coupling relationships among multiple tasks, as well as in the infinite searching spaces of the variational functions. These require a practical and efficient method to jointly optimize NOMA transceivers. In this chapter, we resort to DL in solving this very complicated optimization problem, which is shown to achieve higher performance than conventional variational optimization methods, e.g. mean-field-based methods [40]. Our general idea is to parameterize the variational functions with DNNs and conduct the data-driven end-to-end training. According to the universal approximation theorem of DNN [41], we have the following Theorem 6.1.

Theorem 6.1 (Universal Approximation Theorem for DNN in Approximating Optimal NOMA Transceivers) *The optimal NOMA transceivers, i.e., $\left[f_i^*\right]_{i=1}^{N}$ and $\left[g_i^*\right]_{i=1}^{N}$, which maximize $\mathcal{P}1$, can be accurately approximated by DNNs in the following sense: $\forall \epsilon > 0$, there exists a set of DNNs, i.e., $\left[f_i^{DNN}\right]_{i=1}^{N}$ and $\left[g_i^{DNN}\right]_{i=1}^{N}$, such that*

$$\left| \mathcal{L}\left(\left[f_i^{DNN}\right]_{i=1}^{N}, \left[g_i^{DNN}\right]_{i=1}^{N} \right) - \mathcal{L}\left(\left[f_i^*\right]_{i=1}^{N}, \left[g_i^*\right]_{i=1}^{N} \right) \right| < \epsilon. \tag{6.9}$$

Proof Please refer to Appendix 1. □

Remark 6.1 Theorem 6.1 ensures that the optimal transceivers can be accurately represented by certain DNNs. However, the algorithmic learnability of these DNNs is not assured.

To efficiently learn good NOMA transceivers, elaborate designs on both network structure and learning algorithm are required [6, 42], as we will discuss in the following sections.

6.3 DeepNOMA: An End-to-End DL Framework for NOMA Based on Multi-task Learning

This section proposes an end-to-end DL framework for NOMA, namely Deep-NOMA. Deep multi-task learning is exploited to simplify the network structure without sacrificing the expression ability of the model. An end-to-end training algorithm is also proposed based on an elaborately designed multi-task balancing technique.

6.3.1 Deep Multi-task Learning

Different from conventional DL methods which optimize a single and specific task, deep multi-task learning realizes inductive migration among multiple associated tasks [43]. With the shared layers, one task may provide inductive bias to other tasks which encourages the parameters to converge with better generalization. Recently, deep multi-task learning has been widely applied in machine learning community to achieve a good collaboration among tasks [43].

6.3.2 Network Structure of DeepNOMA

While conventional methods regard NOMA as a single learning task [33, 37], DeepNOMA takes the advantage of the inherent multi-task structure of NOMA. As depicted in Fig. 6.1, DeepNOMA is designed in a task-specific fashion, and is composed of a signature mapping module, namely DeepMAS, a channel module, and a multi-user detection module, namely DeepMUD. In the forward-propagation phase, source messages first flow through DeepMAS to derive the multi-dimensional complex symbols, and then the symbols are superimposed through a channel module given the channel coefficients. Finally, the superimposed signal is decoupled to accurately recover source messages based on task-specific sub-networks in DeepMUD. Performance indicators, such as peak-to-average ratio and activation error rate [28], can be incorporated in DeepNOMA by designing the corresponding sub-networks in DeepMUD.

To train DeepNOMA, the network structures should be specified, and the dataset as well as the loss function should be properly organized. We detail these in the following.

6.3.2.1 DeepMAS

As the input of DeepMAS, we employ one-hot encoding to represent the source messages [6], i.e., each message $\mathcal{M}_n \in \{1, \cdots, M\}$ is represented by a M-dimensional one-hot vector $\mathbf{m}_n = [\mathfrak{m}_{n,m}]_{m=1}^M \in \{0, 1\}^M$ with the $\mathcal{M}_n$-th element equals to 1 and the others are 0.

Note that NOMA users adopt a distributed fashion to map source messages to the transmit signals. Hence, we deploy N disconnected MAS mapping DNNs in DeepMAS, denoted as $f_{\boldsymbol{\mathcal{W}}_n}^{\mathrm{DNN}} : \mathbf{m}_n \to \mathbf{x}_n^\dagger \in \mathbb{C}^{K'}$, $K' \leq K$, $n = 1 \cdots N$, where $\boldsymbol{\mathcal{W}}_n$ is the parameter set. The term $f_{\boldsymbol{\mathcal{W}}_n}^{\mathrm{DNN}}$ can take any form of feed-forward DNNs, such as fully-connected DNN (FC-DNN) and convolutional NN (CNN). If a FC-DNN with L_T layers is assumed, then $f_{\boldsymbol{\mathcal{W}}_n}^{\mathrm{DNN}}(\cdot)$ is given by

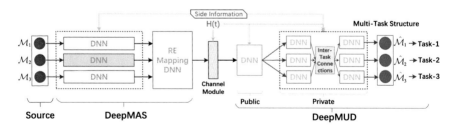

Fig. 6.1 General architecture of the proposed DeepNOMA framework using multi-task DNN

$$\mathbf{x}_n^{\dagger} = f_{\mathcal{W}_n}^{\mathrm{DNN}} (\mathbf{m}_n) = \sigma_f^{(L_T)} \left(\mathbf{W}_n^{(L_T)} \left(\sigma_f^{(L_T-1)} \right. \right.$$
$$\left. \left. \cdots \sigma_f^{(1)} \left(\mathbf{W}_n^{(1)} \mathbf{m}_n + \mathbf{b}_n^{(1)} \right) \cdots \mathbf{b}_n^{(L_T-1)} \right) + \mathbf{b}_n^{(L_T)} \right), \tag{6.10}$$

where $\mathbf{W}_n^{(l)} \in \mathbb{R}^{S_l \times S_{l+1}}$, $l = 1 \cdots L_T$, denotes the parameter matrix associated with layer-l and -$(l+1)$, $\mathbf{b}_n^{(l)}$ denotes the bias associated with layer-$(l+1)$, S_l denotes the number of neurons in layer-l, and $\mathcal{W}_n = \left\{ \mathbf{W}_n^{(1)}, \mathbf{b}_n^{(1)}, \cdots, \mathbf{W}_n^{(L_T)}, \mathbf{b}_n^{(L_T)} \right\}$. The term $\sigma^{(l)}(\cdot)$ corresponds to the activation function of layer-l, which shall be selected according to the purpose of this layer [3]. A normalization operation is employed afterwards to normalize the transmit energy. Normally, (6.10) leads to non-linear MAS mapping similar to SCMA [14], and its linear variant can be obtained by setting $L_T = 1$ and using identity activation function.

To make the proposed framework generalize to both sparse and dense NOMA schemes, an RE mapping layer is introduced to bring sparsity, if needed, to the multi-dimensional symbol $\mathbf{x}_n^{\dagger}$. The output of this layer is given by $\mathbf{x}_n = \mathbf{B}_n \mathbf{x}_n^{\dagger}$, where $\mathbf{B}_n \in \{0, 1\}^{K \times K'}$ is a binary matrix, and the sum of each column in $\mathbf{B}_n$ equals to 1. Ultimately, $f_n(\cdot)$ in (6.1) is parameterized as

$$f_n(\cdot) \leftarrow \mathbf{B}_n f_{\mathcal{W}_n}^{\mathrm{DNN}}(\cdot), \tag{6.11}$$

where $\mathbf{B}_n$ can be either alterable during training or fixed beforehand by reusing the RE mapping patterns optimized for different NOMA schemes [44]. If not specified, the rest of this chapter will assume a fixed and simple RE mapping where each user can get access to all REs, i.e., $K = K'$ and $\mathbf{B}_n = \mathbf{I}$ [13, 16].

The N parallel outputs of DeepMAS are then superimposed in the channel module as follows

$$\mathbf{y} = \sum_{n=1}^{N} \sqrt{P_n} \, \mathrm{diag} \, (\mathbf{h}_n) \, f_{\mathcal{W}_n}^{\mathrm{DNN}}(\mathbf{m}_n) + \mathbf{n}, \tag{6.12}$$

where $\mathbf{h}_n$ is randomly generated according to a block fading channel model. This chapter assumes fixed power allocation where each user transmits its signal with the maximum possible power, i.e., P_n. In Sect. 6.6, we will observe that the power gain differences among the users on each RE can be automatically learned by Deep-NOMA.

6.3.2.2 DeepMUD

While miscellaneous tasks can be incorporated into DeepMUD, this chapter will mainly focus on the message recovery tasks given $\mathbf{y}$. Here we assume that the channel coefficients are perfectly known at the receiver. To cope with N transmission tasks, DeepMUD consists of a public part and N task-specific private parts, as shown in Fig. 6.1. The public part firstly converts the highly-condensed received signal to a high-dimensional feature vector shared by all tasks, for the ease of signal decomposition. With the converted feature vector, the private part is able to deploy a relatively simple structure to extract the private feature vector of each task for message recovery. Without the public part, each private part shall respectively conduct the feature conversion without the knowledge provided by other tasks. This certainly increases the network complexity and makes the learning much harder.

Specifically, the public part of DeepMUD aims to extract the shared feature vector $\mathbf{a}_{\text{Pub}}$

$$\mathbf{a}_{\text{Pub}} = g_{\boldsymbol{\mathcal{V}}_{\text{Pub}}}^{\text{DNN}}(\mathbf{y}; [\mathbf{h}_n]_{n=1}^N), \tag{6.13}$$

where $\boldsymbol{\mathcal{V}}_{\text{Pub}}$ represents the parameter set of the public part. Then $\mathbf{a}_{\text{Pub}}$ is fed into the private parts of DeepMUD to generate the probability vector of the estimated messages, denoted as $\mathbf{a}_n = [a_{n,m}]_{m=1}^M \in (0, 1)^M$, $n = 1 \cdots N$, which is given by

$$\mathbf{a}_n = g_{\boldsymbol{\mathcal{V}}_{\text{Pri-}n}}^{\text{DNN}}(\mathbf{a}_{\text{Pub}}; [\mathbf{h}_n]_{n=1}^N), \tag{6.14}$$

where $\boldsymbol{\mathcal{V}}_{\text{Pri-}n}$ is the parameter set of the n-th private part and $a_{n,m}$ represents the prediction probability of the estimation of $\mathcal{M}_n$. We note that a softmax activation function is deployed at the last layer of $g_{\boldsymbol{\mathcal{V}}_{\text{Pri-}n}}^{\text{DNN}}$ to ensure $\|\mathbf{a}_n\|_1 = 1$. Finally, $g_n(\cdot)$ in (6.3) is parameterized as

$$g_n(\cdot) \leftarrow g_{\boldsymbol{\mathcal{V}}_n}^{\text{DNN}}(\cdot) = g_{\boldsymbol{\mathcal{V}}_{\text{Pri-}n}}^{\text{DNN}} \circ g_{\boldsymbol{\mathcal{V}}_{\text{Pub}}}^{\text{DNN}}(\cdot), \tag{6.15}$$

where $\circ$ is the composition operator, $g_{\boldsymbol{\mathcal{V}}_n}^{\text{DNN}}(\cdot)$ is the detection network for the n-th task and $\mathbf{V}_n = \{\mathbf{V}_{\text{Pub}}, \mathbf{V}_{\text{Pri-}n}\}$ is the parameter set. Similar to (6.10), $g_{\boldsymbol{\mathcal{V}}_{\text{Pub}}}^{\text{DNN}}(\cdot)$ and $g_{\boldsymbol{\mathcal{V}}_{\text{Pri-}n}}^{\text{DNN}}(\cdot)$ can simply take the form of FC-DNN. More sophisticated network designs are considered in Sects. 6.4 and 6.5 to take advantage of the underlaying signal structure of NOMA.

6.3.2.3 Dataset and Loss Function Design

Given the network structure of DeepNOMA, the training dataset and the loss function are two remaining design aspects to achieve end-to-end optimization. Denote $\mathbf{m}_n^{(t)}$, $t = 1 \cdots T$, as the t-th i.i.d. one-hot samples generated according to the prior distribution of source messages. The synthetic dataset $\mathfrak{D}$ is given by

$$\mathfrak{D} = \left\{ \left[\mathbf{m}_1^{(t)}, \cdots, \mathbf{m}_n^{(t)}, \cdots, \mathbf{m}_N^{(t)} \right] \right\}_{t=1}^T = \left\{ \mathfrak{M}^{(t)} \right\}_{t=1}^T, \tag{6.16}$$

where T is the size of the dataset. Since DeepNOMA aims to recover the transmit messages at the receiver, $\mathfrak{D}$ is applied as both the input data and the output label.

Observing that NOMA detection problem is to recover the source messages in a limited search space, the problem is equivalent to a typical classification problem in the machine learning field. This fact motivates us to employ the widely used cross-entropy (CE) loss function to train DeepNOMA [3]. CE loss takes the form

$$H(p, q) = -\mathbb{E}_{p(x)}[\log q(x)], \qquad (6.17)$$

which measures the difference between the true probability distribution $p(x)$ and the estimated distribution $q(x)$. Recall the objective function (6.8) of the overall optimization problem $\mathcal{P}1$: each additive component item in (6.8), i.e., $\mathcal{L}([f_i]_{i=1}^N, g_n)$, takes the form of negative CE where $p(x)$ and $q(x)$ correspond to $P(\mathbf{y}|\mathcal{M}; [f_i]_{i=1}^N)$ and $P(\mathcal{M}_n|\mathbf{y}; g_n)$, respectively.

By replacing $\mathcal{M}$ with one-hot representation $\mathfrak{M}^{(t)}$ in (6.16), we can derive the negative CE reconstruction loss of the n-th transmission task over $\mathfrak{D}$, as follows,

$$\mathcal{L}_n^{CE}\left([f_{\mathbf{W}_i}^{DNN}]_{i=1}^N, g_{\mathbf{V}_n}^{DNN}; \mathfrak{D}\right)$$
$$= \sum_{t=1}^T \mathbb{E}_{P\left(\mathbf{y}|\mathfrak{M}^{(t)}; [f_{\mathbf{W}_i}^{DNN}]_{i=1}^N\right)}\left[\log P\left(\mathbf{m}_n^{(t)}|\mathbf{y}; g_{\mathbf{V}_n}^{DNN}\right)\right], \qquad (6.18)$$

where the expatiation over $P(\mathcal{M})$ in (6.6) is replaced by the empirical expectation over $\mathfrak{D}$. Aggregating (6.8) and (6.18), the objective function of $\mathcal{P}1$ can also be converted into a negative CE loss over $\mathfrak{D}$ as follows

$$\mathcal{L}^{CE}\left([f_{\mathbf{W}_n}^{DNN}]_{n=1}^N, [g_{\mathbf{V}_n}^{DNN}]_{n=1}^N; \mathfrak{D}\right)$$
$$= \sum_{n=1}^N w_n \mathcal{L}_n^{CE}\left([f_{\mathbf{W}_i}^{DNN}]_{i=1}^N, g_{\mathbf{V}_n}^{DNN}; \mathfrak{D}\right)$$
$$= \sum_{t=1}^T \sum_{n=1}^N w_n \mathbb{E}_{P\left(\mathbf{y}|\mathfrak{M}^{(t)}; [f_{\mathbf{W}_i}^{DNN}]_{i=1}^N\right)}\left[\log P\left(\mathbf{m}_n^{(t)}|\mathbf{y}; g_{\mathbf{V}_n}^{DNN}\right)\right] \qquad (6.19)$$
$$\overset{(a)}{\approx} \sum_{t=1}^T \sum_{n=1}^N w_n \sum_{s=1}^S \log P\left(\mathbf{m}_n^{(t)}\right|$$
$$\sum_{i=1}^N \sqrt{P_i} \operatorname{diag}(\mathbf{h}_i) f_{\mathbf{W}_i}^{DNN}\left(\mathbf{m}_i^{(t)}\right) + \boldsymbol{\epsilon}^{(s)}; g_{\mathbf{V}_n}^{DNN}\right),$$

where step-(a) applies the reparameterization trick to estimate the expectation over the posterior distribution of $\mathbf{y}$ and $\boldsymbol{\epsilon}^{(s)} \sim \mathcal{CN}(\mathbf{0}, \sigma_0^2 \mathbf{I})$ [28], and σ_0^2 is the noise variance during training. The log term in (6.19) can be further simplified as [40]

$$\log P\left(\mathbf{m}_n^{(t)}\bigg|\sum_{i=1}^{N}\sqrt{P_i}\,\mathrm{diag}\,(\mathbf{h}_i)\,f_{\boldsymbol{\mathcal{W}}_i}^{\mathrm{DNN}}\left(\mathbf{m}_i^{(t)}\right)+\boldsymbol{\epsilon}^{(s)};\,g_{\boldsymbol{\mathcal{V}}_n}^{\mathrm{DNN}}\right)$$

$$= \mathbf{m}_n^{(t)\top}\log\left(\mathbf{a}_n^{(t)}\right)=\sum_{m=1}^{M}\mathrm{m}_{n,m}^{(t)}\log\left(a_{n,m}^{(t)}\right),$$

(6.20)

where

$$\mathbf{a}_n^{(t)}=g_{\boldsymbol{\mathcal{V}}_n}^{\mathrm{DNN}}\left(\sum_{i=1}^{N}\sqrt{P_i}\,\mathrm{diag}\,(\mathbf{h}_i)\,f_{\boldsymbol{\mathcal{W}}_i}^{\mathrm{DNN}}\left(\mathbf{m}_i^{(t)}\right)+\boldsymbol{\epsilon}^{(s)}\right).$$

(6.21)

Substitute (6.20) into (6.19) and we derive an equivalent optimization problem of $\mathcal{P}1$ with DeepNOMA framework as follows

$$\mathcal{P}2: \min_{\boldsymbol{\mathcal{W}}_n,\boldsymbol{\mathcal{V}}_n,n=1\cdots N}\,-\mathcal{L}^{\mathrm{CE}}\left(\left[f_{\boldsymbol{\mathcal{W}}_n}^{\mathrm{DNN}}\right]_{n=1}^{N},\left[g_{\boldsymbol{\mathcal{V}}_n}^{\mathrm{DNN}}\right]_{n=1}^{N};\,\mathfrak{D}\right).$$

(6.22)

Gradient-based optimization methods are then applied to solve $\mathcal{P}2$ which can approach global minimum with high probability after enough iterations. Without ambiguity, we omit the input arguments of (6.18) and (6.19) in the following, and use $\mathcal{L}_n^{\mathrm{CE}}$ and $\mathcal{L}^{\mathrm{CE}}$, respectively, for brevity.

Now we analyze the utility of the multi-task structure in DeepNOMA. First of all, the joint representation of all tasks with FC-DNN suffers from the curse of dimensionality, i.e., a very large DNN is required, which is hard to train. Using a multi-task structure enables that each task can be dealt with a relatively small DNN. Secondly, the public parts and the private parts in DeepMUD promote the level-by-level extraction of useful features, which enhances the learning performance. Last but not least, the multi-task structure facilitates the introduction of interference-cancellation as the inter-task connection, as we will expatiate in Sect. 6.5.

6.3.3   Multi-task Balancing Technique

In NOMA, multiple transmission tasks ought to be simultaneously optimized. However, during training it is occasionally that some tasks are prioritized and some are neglected due to the random fluctuation of gradient-based optimizer, which causes the unfairness among various tasks. Therefore, the tasks involved in $\mathcal{P}2$ should be properly balanced during training so that the parameters can converge to the status which benefits all tasks.

In this chapter, we develop a multi-task balancing technique to guarantee fairness among the users and to avoid getting stuck in local optima. The core idea is to weaken or reinforce the backpropagated gradients of the tasks whose performances are beyond or below average. To achieve this, we penalize $\mathcal{P}2$ with an additional task balancing loss

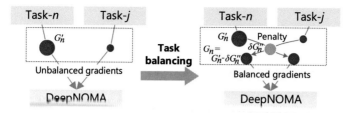

Fig. 6.2 Effect of multi-task balancing in DeepNOMA

$$\mathcal{L}^{B} = \mathcal{D}\left(\frac{1}{N}\left[\mathcal{L}^{CE}, \cdots, \mathcal{L}^{CE} \right], \left[\mathcal{L}_{1}^{CE}, \cdots, \mathcal{L}_{N}^{CE} \right] \right), \qquad (6.23)$$

where $\mathcal{D}(\mathcal{A}, \mathcal{B})$ is a predefined distance measurement between vectors $\mathcal{A} = [A(n)]_{n=1}^{N}$ and $\mathcal{B} = [B(n)]_{n=1}^{N}$. Here we apply Minkowski distance of order p as the distance measurement, i.e., $\mathcal{D}(\mathcal{A}, \mathcal{B}) = \left(\sum_{n=1}^{N} |A(n) - B(n)|^{p} \right)^{1/p}$.

Finally, DeepNOMA is optimized by minimizing the following problem

$$\mathcal{P}3 : \min_{\mathbf{W}_n, \mathbf{V}_n, n=1\cdots N} -\mathcal{L}\left(\left[f_{\mathbf{W}_n}^{DNN} \right]_{n=1}^{N}, \left[g_{\mathbf{V}_n}^{DNN} \right]_{n=1}^{N} ; \mathfrak{D} \right), \qquad (6.24)$$

$$\text{s.t. } \mathcal{L} = \mathcal{L}^{CE} + b\mathcal{L}^{B}, \qquad (6.25)$$

where b is a hyper-parameter related to the task balancing term. With such penalty, the n-th task, which has gradient G'_n with high training speed according to $\mathcal{P}2$, tends to be slowed down by a refining term $\delta G''_n$, as shown in Fig. 6.2.

Furthermore, we note that, introducing the task balancing loss can be regarded as a generalization of min-max optimization. To see this, we consider a simple two-task case where $N = 2$, $b = 1$, $p = 1$, and $w_n = 1$. Denote $-\mathcal{L}^{(j)}$ as the loss in the j-th training epoch. Without loss of generality, we assume that $-\mathcal{L}_2^{CE,(j)} > -\mathcal{L}_1^{CE,(j)}$, which reflects that the first task has better performance in the previous training. The overall loss $-\mathcal{L}^{(j)}$ is then given by

$$\begin{aligned} -\mathcal{L}^{(j)} &= -\mathcal{L}^{CE,(j)} - \sum_{n=1}^{2} \left| \frac{1}{2}\mathcal{L}^{CE,(j)} - \mathcal{L}_n^{CE,(j)} \right| \\ &= -\sum_{n=1}^{2} \mathcal{L}_n^{CE,(j)} + \mathcal{L}_1^{CE,(j)} - \mathcal{L}_2^{CE,(j)} = -2\mathcal{L}_2^{CE,(j)}, \end{aligned} \qquad (6.26)$$

Substitute (6.26) into (6.25), and $\mathcal{P}3$ is converted to an equivalent min-max problem

$$\mathcal{P}3' : \min_{\mathbf{W}_n, \mathbf{V}_n, n=1,2} \max_{n} -\mathcal{L}_n^{CE}, \qquad (6.27)$$

where only the task with poor performance is optimized for each epoch.

Algorithm 6.1 Training Algorithm of DeepNOMA

Require: System settings "N", "M", "K", "σ_0^2", "w_n", "$\mathfrak{D}$", "S", "b", and "p".

Ensure: Network parameter sets $\boldsymbol{W}_n^{(j)}$ and $\boldsymbol{V}_n^{(j)}$ of DeepMAS and DeepMUD, respectively.

1: $\boldsymbol{W}_n^{(0)}, \boldsymbol{V}_n^{(0)}, i = 1 \cdots N$ $\leftarrow$ Initialize network parameter sets

2: **repeat**

3:     *(Forward propagation)*

4:     $j \leftarrow 1$

5:     $\mathfrak{D}' = \{\mathfrak{M}^{(t)}\}_{t=1}^{T'}$ Draw a batch of data with size T' out of $\mathfrak{D}$

6:     $\mathcal{L}_{\text{Batch}}^{(j)} \leftarrow \mathcal{L}\left(\left[f_{\boldsymbol{W}_n}^{\text{DNN}} \right]_{n=1}^N, \left[g_{\boldsymbol{V}_n}^{\text{DNN}} \right]_{n=1}^N ; \mathfrak{D}' \right)$ in (6.25)

7:     *(Backward propagation)*

8:     $\nabla_{\boldsymbol{W}_n}, \nabla_{\boldsymbol{V}_n} \leftarrow \nabla_{\boldsymbol{W}_n, \boldsymbol{V}_n} \mathcal{L}_{\text{Batch}}^{(j)}, n = 1 \cdots N$

9:     $\boldsymbol{W}_n^{(j)}, \boldsymbol{V}_n^{(j)} \leftarrow$ Update network parameters based on $\boldsymbol{W}_n^{(j-1)}, \boldsymbol{V}_n^{(j-1)}$ and $\nabla_{\boldsymbol{W}_n}, \nabla_{\boldsymbol{V}_n}$ with gradient-based optimizer, $n = 1 \cdots N$

10:     $j \leftarrow j+1$

11: **until** Convergence of $\boldsymbol{W}_n^{(j)}$ and $\boldsymbol{V}_n^{(j)}$

6.3.4   Training Algorithm

This chapter adopts the offline-training and online-deploying mode [6]. We first train DeepNOMA in offline mode to derive good transceivers. Then the trained DeepMAS and DeepMUD are respectively deployed at the transmitters and the receiver in online mode for NOMA transmissions. In offline training phase, we apply forward- and backward-propagations to conduct end-to-end optimization based on $\mathcal{P}3$, as illustrated in Algorithm 6.1. For each training epoch, a batch of data $\mathfrak{D}'$ is randomly drawn out of $\mathfrak{D}$. Then gradient-based optimizer, such as Adam, is employed to update $\boldsymbol{W}_n$ and $\boldsymbol{V}_n$ given certain network structures of DeepNOMA.

Finally, we note that, while DeepNOMA framework is proposed in the uplink, it is straightforward to extend it to the downlink with two minor modifications: (1). Deploy the entire DeepMAS at the BS; and (2). Deploy the user-specific DeepMUD at each user.

6.4   DeepMAS: Model-Based MAS Mapping Network Design

This section focuses on the network structure and the parameter initialization of DeepMAS, where insightful communication model is incorporated to reduce implementation complexity.

6.4.1 Model-Based Transmitter Design

Given a certain network structure of DeepNOMA, we can deploy Algorithm 1 and generate good MAS mappings for NOMA. If DeepMAS is parameterized by FC-DNN, the trained MAS mappings may not have irregular shapes, i.e., the transmit signal alphabet on each RE does not have regular geometric shapes, unlike the conventional QAM constellations. However, for ease of hardware implementation, transmit signal alphabets are required to be constrained to certain shapes [12]. Hence, we introduce shape prior into the network structure of DeepMAS.

We illustrate the proposed model-based DeepMAS in Fig. 6.3. Given parameter set $\mathcal{W}_n$ of the n-th user, a pre-defined MAS mapping model, namely $\mathcal{R}(\cdot)$, is employed to map $\mathcal{W}_n$ to a set of complex symbol sequences

$$\mathcal{R} : \mathcal{W}_n \rightarrow \mathcal{X}_n^{\dagger} = \left[\mathbf{x}_{n,1}^{\dagger}, \cdots, \mathbf{x}_{n,M}^{\dagger} \right], \mathbf{x}_{n,m}^{\dagger} \subset \mathbb{C}^{K'}. \tag{6.28}$$

where $\mathcal{X}_n^{\dagger}$ is called the MAS pool for the n-th user. With one-hot input $\mathbf{m}_n$, one element is selected out of $\mathcal{X}_n^{\dagger}$ and then mapped to REs for NOMA transmission, as follows

$$\mathbf{x}_n = \mathbf{B}_n \mathbf{x}_n^{\dagger} = \mathbf{B}_n \mathcal{R} \left(\mathcal{W}_n \right) \mathbf{m}_n. \tag{6.29}$$

Here, $\mathcal{R}(\cdot)$ implies the network structure of DeepMAS and can be set as typical constellation shape. During end-to-end training, only $\mathcal{W}_n$ is optimized.

As an example, we consider a case where the complex symbol on each dimension of MAS has the shape of parallelogram. Hence, the parameter set of the n-th user can be defined as $\mathcal{W}_n = \{L_n, S_n, \theta_{L,n}, \theta_{S,n}\}$, where L_n is the length of the longer side, S_n is the length of the shorter side, and $\theta_{L,n}$ and $\theta_{L,n}$ are respectively defined as the intersection angles related to the longer and the shorter sides. Further, if we set $K = 4$ and $K' = 2$, which is a typical setting of sparse NOMA with 50% sparsity, $\mathcal{R}(\cdot)$ can be expressed by

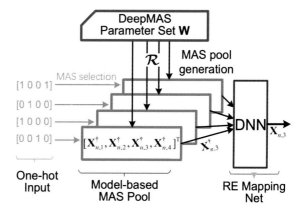

Fig. 6.3 Model-based network structure for DeepMAS

$$\mathcal{R}\left(\boldsymbol{W}_n\right) =$$

$$\begin{bmatrix} L_n \cos(\theta_{L,n}) + j S_n \cos(\theta_{S,n}), \, L_n \sin(\theta_{L,n}) + j S_n \sin(\theta_{S,n}) \\ -L_n \cos(\theta_{L,n}) + j S_n \cos(\theta_{S,n}), -L_n \sin(\theta_{L,n}) + j S_n \sin(\theta_{S,n}) \\ L_n \cos(\theta_{L,n}) - j S_n \cos(\theta_{S,n}), \, L_n \sin(\theta_{L,n}) - j S_n \sin(\theta_{S,n}) \\ -L_n \cos(\theta_{L,n}) - j S_n \cos(\theta_{S,n}), -L_n \sin(\theta_{L,n}) - j S_n \sin(\theta_{S,n}) \end{bmatrix}^{\top} . \tag{6.30}$$

6.4.2 Parameter Initialization

We now analyze the parameter initialization method which may affect the network training performance [3]. Motivated by the pre-training technique which aims to initialize the parameters in a basin close to an apparent local minimum associated with better generalization [45], we propose to initialize the network parameters such that the initialized composite signal of DeepMAS follows the optimum distribution in an information-theoretic perspective.

Assuming ML detection, optimal MAS mapping is derived by maximizing the mutual information (MI) between the receiver and transmitters, i.e., $\max I(\mathbf{x}_1, \mathbf{x}_2, \cdots, \mathbf{x}_N; \mathbf{y})$. Denote $h(\cdot)$ as the differential entropy function, and we have

$$I(\mathbf{x}_1, \mathbf{x}_2, \cdots, \mathbf{x}_N; \mathbf{y}) = h(\mathbf{y}) - h(\mathbf{n}). \tag{6.31}$$

Since $h(\mathbf{n})$ is a constant, maximizing MI can be equivalently converted to maximizing the entropy of the random vector $\mathbf{y}$ [20], i.e.,

$$\mathcal{P}4: \max_{f_n, n=1\cdots N} h(\mathbf{y}), \text{ s.t. } (6.1) \text{ and } (6.2). \tag{6.32}$$

According to $\mathcal{P}4$, the parameters shall be initialized to enforce maximum entropy. To simplify (6.32), we define the non-Gaussianity (NG) of $\mathbf{y}$ as the KL divergence between the probability distribution of $\mathbf{y}$ and a normal distribution with the same mean and co-variance as $\mathbf{y}$

$$NG(\mathbf{y}) = KL\left(P(\mathbf{y}), \mathcal{N}\left(\boldsymbol{\mu}_\mathbf{y}, \boldsymbol{\Sigma}_\mathbf{y}\right)\right) \geq 0, \tag{6.33}$$

where $\boldsymbol{\mu}_\mathbf{y}$ and $\boldsymbol{\Sigma}_\mathbf{y}$ are the mean and co-variance of $\mathbf{y}$, respectively. Then we have the following result.

Theorem 6.2 *Optimal composite MAS mappings, as defined in $\mathcal{P}4$, can be equivalently derived by solving*

$$\mathcal{P}5: \min_{f_n, n=1\cdots N} NG(\mathbf{y}) - h^U(\mathbf{y}), \text{ s.t. } (6.1) \text{and } (7.1), \tag{6.34}$$

where $h^U(\mathbf{y}) = K \log(2\pi e) + \frac{1}{2}\left(\left|\left(\boldsymbol{\Sigma}_\mathbf{y}\right)\right|\right)$ is an upper bound of $h(\mathbf{y})$ derived by Gaussian approximation.

Proof Please refer to Appendix 2. □

Remark 6.2 $h^{U}(\mathbf{y})$ represents the total power of the composite signal, and thus is approximately a constant value with a large number of users according to the law of large numbers.

Therefore, Theorem 6.2 indicates that better MI can be achieved by minimizing NG($\mathbf{y}$). To this end, our initialization principle is to randomly generate the parameters such that the initialized composite signal of DeepMAS approaches a Gaussian distribution, i.e., $\min_{\mathbf{W}_n^{(0)}}$ NG ($\mathbf{y}$), where $\mathbf{W}_n^{(0)}$ is the initial parameters. For example, if DeepMAS is parameterized by FC-DNN with Sigmoid activation, Xaiver initialization which ensures Gaussian-distributed outputs can be deployed [45]. For the example of model-based DeepMAS in (6.30), we initialize the transaction angles and the sides with uniform and Gaussian distributions, respectively.

6.5 DeepMUD: Interference Cancellation-Based MUD Network Design

By averaging (6.25) over all possible $\mathbf{h}_n$ and deploying Algorithm 6.1, it seems easy to obtain good NOMA transceivers under fading channels. However, deploying a straightforward network structure in DeepMUD will not work well. The major obstacle lays in the very complicated signal space of $\mathbf{y}$ caused by different fading states of users. To tackle the challenge of NOMA detection in fading channels, we propose a sophisticated design of DeepMUD, where the superposition signal structure of $\mathbf{y}$ is exploited by introducing IC-enabled DNN units as inter-task interactions. The representational power of IC-enabled DNN is also analyzed.

6.5.1 Interference Cancellation for Multiple Access Channel

IC is originally developed to achieve the corner points of MAC capacity region with low complexity [10]. The benefits of IC can be illustrated as follows. Suppose that two transmitters wish to communicate two independent messages $\mathcal{M}_1$ and $\mathcal{M}_2$ to the receiver, respectively, over the MAC ($\mathcal{X}_1 \times \mathcal{X}_2, p(y|x_1, x_2), \mathcal{Y}$), where $\mathcal{X}_1$ and $\mathcal{X}_2$ are the input alphabets, $p(y|x_1, x_2)$ represents the symbol-by-symbol superposition operation (such as addition) of MAC, and $\mathcal{Y}$ is the output alphabet. The message pair ($\mathcal{M}_1, \mathcal{M}_2$) is uniform over $[1 : 2^{nR_1}] \times [1 : 2^{nR_2}]$, and ($R_1, R_2$) is the rate tuple. We define two encoders, namely Encoder-1 and Encoder-2, which assign length-n symbol sequences x_1^n and x_2^n to $\mathcal{M}_1$ and $\mathcal{M}_2$, respectively. The decoder assigns an estimated ($\hat{\mathcal{M}}_1, \hat{\mathcal{M}}_2$) to each received sequence y^n. Further, we assume that (R_1, R_2) is set at the corner point of MAC capacity region such that M_1 can be perfectly recovered by regarding interference as noise. We visualize the received signal space

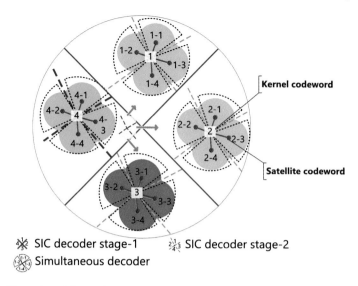

SIC decoder stage-1   SIC decoder stage-2
Simultaneous decoder

Fig. 6.4 Illustration of the received composite codeword space of two-user MAC. All clouds have identical inner structures

in Fig. 6.4 with $|\mathcal{X}_1| = |\mathcal{X}_2| = 4$, where each codeword of Encoder-1, i.e., the kernel of each cloud, is surrounded by satellite codebooks generated by Encoder-2, and all clouds are identical except their kernel positions.

Simultaneous decoding (SD) and SIC have been developed to recover the source messages in MAC [10]. As illustrated by the black-dotted curves in Fig. 6.4, SD jointly recovers the source message pair by dividing the entire codeword space into 16 parts. Hence, SD holds exponential complexity with respect to the number of users. Alternatively, SIC decoder exploits the superposition structure of y^n by firstly identifying the kernel codeword, as shown by the red-solid curve in Fig. 6.4, and secondly deciding the satellite codeword around the kernel, as illustrated by the blue-dashed curve. Thus, SIC holds linear complexity. The complexity reduction of SIC is due to the homogeneity among the clouds, thus the same decoder structure can be reused after SIC.

6.5.2   ICNN: Interference Cancellation-Enabled DNN

Observing the gain of IC, we propose an IC-enabled DNN based on multi-task DNN, namely ICNN, to enhance the representational power in processing superimposed signals. Figure 6.5 presents the conceptional structures of FC-DNN, multi-task DNN and ICNN. To recover the source messages from superimposed inputs, FC-DNN shall construct a joint classifier for both Encoders which causes high classification complexity. Directly employing multi-task structure does not reduce the network

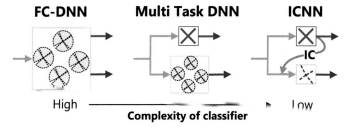

Fig. 6.5 Comparisons among FC-DNN, multi-task DNN and ICNN. DNN works as a classifier to recover source messages by subdividing the input signal space into convex polytopes, as illustrated by the lines or dashes within each block, where ICNN requires the lowest complexity

complexity, since x_2^n is hided in y^n and a simultaneous classification is still required. By explicitly introducing IC as inter-task interactions in ICNN, x_2^n can be dug out and recovered with very simple classifier, as visualized in Fig. 6.5.

Now we formally propose Theorem 6.3 which gives comprehensive evidence that introducing IC provides higher representational power. Define the detection error probability as $P_e^{(n)} = \Pr\{(\hat{M}_1, \hat{M}_2) \neq (M_1, M_2)\}$. Set the cardinality of input alphabet as $|\mathcal{X}_1| = |\mathcal{X}_2| = M$, and reuse the same notations in Sect. 6.5.1. We have the following universal approximation theorem for width-bounded DNN in approximating the optimal MUD.

Theorem 6.3 (Universal Approximation Theorem for Width-Bounded ICNN in Approximating the Optimal MUD in Two-User MAC) *Given an input y^n, for any ϵ and a large enough n, there exists an ICNN $\mathscr{A}$, with the network width bounded by $O(nM^n)$, which maps y^n to $(\hat{M}_1, \hat{M}_2)$ such that $P_e^{(n)} < \epsilon$.*

Proof The proof is done by first constructing DNN blocks to recast the joint typicality decoding and then applying joint typicality lemma. Please refer to Appendix 3 for details.

Compared with Theorem 6.1 which simply states the existence of the optimal DNN-based MUD, Theorem 6.3 provides a bound on the width and depth of this DNN.

Remark 6.3 If FC-DNN is deployed, the width bound scales from $O(nM^n)$ to $O(nM^{2n})$.

Assuming $n = 1$ and $M = 4$, as shown in Fig. 6.4, FC-DNN requires the network size of $O(4^2)$ while ICNN requires $O(4)$. This implies that ICNN achieves higher representational power than FC-DNN with respect to the superimposed input signals.

6.5.3 DeepMUD Based on ICNN

Now we incorporate ICNN as the interplays among multiple tasks in DeepMUD, as illustrated in Fig. 6.6. By doing this, the highly condensed input signal can be decomposed into several less-interfered signal streams, which promotes the learning efficiency. Here, the key design aspect is to enable flexible and learnable IC. We elaborate on the detailed design of ICNN-based DeepMUD as follows. For brevity, we have omitted the notations of parameter set related to DNNs.

Our proposed DeepMUD encompasses of a common preprocessing stage and L IC-based detection stages. First of all, the received signal $\mathbf{y}$ is preprocessed based on a simple equalization regarding interference-as-noise (IAN), which is embedded in

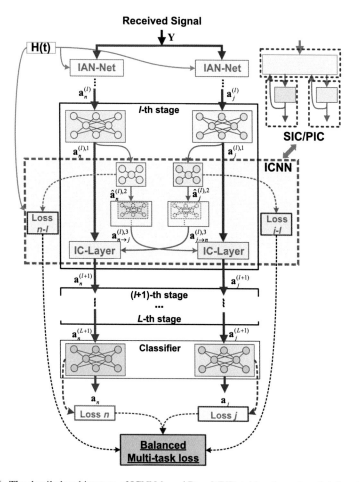

Fig. 6.6 The detailed architecture of ICNN-based DeepMUD, taking the n-th and j-th tasks as an example

IAN-Net block as shown in Fig. 6.6. As an example, we deploy a simple least-square equalization in IAN-Net, denoted as $g_n^{\text{IAN}}(\cdot)$, and the activated feature vector $\mathbf{a}_n^{(1)}$ of the n-th task is derived as

$$\mathbf{a}_n^{(1)} = g_n^{\text{IAN}}(\mathbf{y}; \mathbf{h}_n) = (\text{diag}\,(\mathbf{h}_n))^{-1}\,\mathbf{y}, n = 1 \cdots N, \tag{6.35}$$

where $\mathbf{y}$ and $\mathbf{h}_n$ are taken as the inputs. Note that other preprocessing methods are not precluded.

The preprocessed feature vector $\mathbf{a}_n^{(1)}$ is then propagated to the subsequent detection stages to cancel inter-task interference. We denote the input and output signals of the l-th stage, $l = 1 \cdots L$, of the n-th task as $\mathbf{a}_n^{(l)}$ and $\mathbf{a}_n^{(l+1)}$, respectively; and $\mathbf{a}_n^{(1)}$ in (6.35) is the input of the first stage. The feature mapping function in the l-th stage is written as $g_n^{(l)}(\cdot)$, i.e.,

$$\mathbf{a}_n^{(l+1)} = g_n^{(l)}\left(\mathbf{a}_n^{(l)}\right). \tag{6.36}$$

Finally, $\mathbf{a}_n^{(L+1)}$ is forward-propagated through a classification layer $g_n^{\text{Cls}}(\cdot)$, e.g. a SoftMax layer, to generate the estimated message $\mathbf{a}_n$ of the n-th user

$$\mathbf{a}_n = g_n^{\text{Cls}}\left(\mathbf{a}_n^{(L+1)}\right). \tag{6.37}$$

For each time slot, $\mathbf{a}_n$ is generated and used in (6.20) for loss calculation.

Now we focus on the detailed signal flow in the l-th stage of the n-th task. To begin with, $\mathbf{a}_n^{(l)}$ is forward-propagated through a DNN to extract the hidden representations $\mathbf{a}_n^{(l),1}$ as follows

$$\mathbf{a}_n^{(l),1} = g_n^{(l),1}\left(\mathbf{a}_n^{(l)}\right). \tag{6.38}$$

Then $\mathbf{a}_n^{(l),1}$ is propagated through a DNN-based classifier for a rough and immediate estimation

$$\hat{\mathbf{a}}_n^{(l),2} = g_n^{(l),\text{Cls}}\left(\mathbf{a}_n^{(l),1}\right). \tag{6.39}$$

Here $g_n^{(l),1}(\cdot)$ extracts the shared hidden feature for both the immediate estimation in $g_n^{(l),\text{Cls}}(\cdot)$ and the subsequent processing, i.e., interference cancellation as well as the final detection. The output signal $\hat{\mathbf{a}}_n^{(l),2}$ is an estimated message of the n-th user, which is utilized to estimate the mutual interference caused by NOMA transmissions. For example, we define $\mathbf{a}_{n \to j}^{(l),3}$ as the interference component caused by the n-th task at the l-th stage of the j-th task. $\mathbf{a}_{n \to j}^{(l),3}$ can be obtained as follows

$$\mathbf{a}_{n \to j}^{(l),3} = g_j^{(l),1}\left(g_j^{(l-1)} \cdots \left(g_j^{(1)}\left(g_j^{\text{IAN}}\left(f_n\left(\hat{\mathbf{a}}_n^{(l),2}\right); \mathbf{h}_j\right)\right)\right)\right), \tag{6.40}$$

where $\hat{\mathbf{a}}_n^{(l),2}$ is first encoded with the n-th DeepMAS and then propagated through the previous stages of the j-th task.

For the n-th task, we exploit $\mathbf{a}_{j \to n}^{(l),3}$, $\forall j \neq n$, as the external information to cancel the inter-task interference. To achieve this, we introduce a IC-layer, denoted by $g_n^{(l),\text{IC}}(\cdot)$, as follows

$$\mathbf{a}_n^{(l+1)} = g_n^{(l),\text{IC}}\left(\mathbf{a}_n^{(l),1}; \mathbf{a}_{j\to n}^{(l),3}, \forall j \neq n\right), \tag{6.41}$$

where the interference is partially eliminated from the original interfered signal $\mathbf{a}_n^{(l),1}$ according to the knowledge of interfering signals. By parameterizing $g_n^{(l),\text{IC}}(\cdot)$ with a general DNN, IC-layer is able to conduct non-linear IC. Whereas Gaussian MAC performs linear additions during wireless transmission, we can simply use a linear subtraction at IC-layers as follows

$$g_n^{(l),\text{IC}}\left(\mathbf{a}_n^{(l),1}; \mathbf{a}_{j\to n}^{(l),3}, \forall j \neq n\right) = \mathbf{a}_n^{(l),1} - \sum_{j=1, j\neq n}^{N} \gamma_{n,j}^{(l)} \mathbf{a}_{j\to n}^{(l),3}, \tag{6.42}$$

where $\gamma_{n,j}^{(l)}$ is a learnable scaler factor, namely IC factor, at the j-th stage of the n-th task to control the amplitude of IC. With the automatically learned IC factor, the proposed DeepMUD can achieve a good tradeoff between error propagation and interference cancellation.

After IC-layer, $\mathbf{a}_n^{(l+1)}$ is obtained and then sent into the $(l+1)$-th stage. Finally, the overall expression of $g_n(\cdot)$ in (6.3) is given by

$$g_n(\cdot) = g_n^{\text{Cls}} \circ g_n^{(L)} \cdots \circ g_n^{(l)} \cdots \circ g_n^{(1)}(\cdot), \tag{6.43}$$

where $g_n^{(l)}$ in (6.36) is given by aggregating all mapping functions within the l-th stage

$$g_n^{(l)}(\cdot) = g_n^{(l),\text{IC}} \circ g_n^{(l),\text{Cls}} \circ g_n^{(l),1}(\cdot). \tag{6.44}$$

We note that, even using $L = 1$ would be enough to achieve better performance than conventional MUDs for typical NOMA system settings, as we will show in Sect. 6.6.

6.5.4 Training DeepMUD over Fading Channel

As aforementioned, the core concept of ICNN-based DeepMUD is to perform a rough estimation in each IC stage of each task and to cancel the inter-task interference accordingly. However, inaccurate estimation in each IC stage may lead to error propagation which worsen the detection performance. Therefore, instead of simply optimizing the *global* loss function $\mathcal{L}$ in (6.25), we also fine-tune the classifiers $g_n^{(l),\text{Cls}}$ within each IC stage to optimize the *local* loss functions, as illustrated in Fig. 6.6.

Without loss of generality, we focus on the local loss function in the l-th stage of the n-th task. Here we aim to minimize the CE loss between the estimated message $\hat{\mathbf{a}}_n^{(l),2}$ and $\hat{\mathbf{M}}_n$. We denote $\hat{\mathbf{a}}_n^{(l),2,(t)} = \left[\hat{a}_n^{(l),2,(t)}(m)\right]_{m=1}^{M}$ as the t-th estimation sample according to (6.16). By replacing $\mathbf{a}_n$ with $\hat{\mathbf{a}}_n^{(l),2,(t)}$ in (6.19) and using (6.20), the local CE loss over $\mathfrak{D}$ is given by

$$\mathcal{L}_n^{\text{CE-}l} = -\sum_{t=1}^{T}\sum_{s=1}^{S}\sum_{m=1}^{M} \mathfrak{m}_{n,m}^{(t)} \log\left(\hat{a}_n^{(l),2,(t)}(m)\right). \tag{6.45}$$

Therefore, for each channel realization $\mathbf{h}_n$, the total loss is given by adding (6.25) and (6.45)

$$\mathcal{L}^{\text{Tot}} = \mathcal{L} + \sum_{n=1}^{N}\sum_{l=1}^{I_r} \zeta_{n,l}\mathcal{L}_n^{\text{CE-}l}, \tag{6.46}$$

where $\zeta_{n,l}$ denotes the weight coeffient.

Finally, we average $\mathcal{L}^{\text{Tot}}$ over the distributions of channel coefficients and derive

$$\mathcal{L}^{\text{Fading,Tot}} = \mathbb{E}_{[\mathbf{h}_n]_{n=1}^N}\left[\mathcal{L}^{\text{Tot}}\right] \approx \sum_{b=1}^{B}\left[\mathcal{L}^{\text{Tot}}\left([\mathbf{h}_n^{(b)}]_{n=1}^N\right)\right] \tag{6.47}$$

where $b = 1 \cdots B$ is the index of channel realizations, and $\mathcal{L}^{\text{Tot}}([\mathbf{h}_n^{(b)}]_{n=1}^N)$ denotes the total loss function given the channel realization $[\mathbf{h}_n^{(b)}]_{n=1}^N$. By minimizing $\mathcal{L}^{\text{Fading,Tot}}$ according to Algorithm 6.1, we are able to jointly optimize NOMA transceivers under fading channel model.

6.6 Simulation Results

In this section, we implement the proposed DeepNOMA framework and perform experiments on synthetic datasets. The well-known Tensorflow framework is adopted to train the proposed DeepNOMA. Each complex number is represented by its real and imaginary parts in real number field, for ease of implementation on Tensorflow. Specifically, we use a GPU server with a Intel® Xeon E5-2620 v4 CPU and two Nvidia® GTX-1080Ti GPUs for acceleration on training, while our framework can still be trained and implemented on an ordinary PC. The depth and the width of DNN adopted in this section are empirically determined, such that increasing the sizes cannot further promote the learning performance. The network parameters are initialized following the principle in Sect. 6.4. Meanwhile, link-level Monte-Carlo simulations are also conducted on MATLAB® to validate the performance gains of our scheme over existing NOMA schemes with respect to detection accuracy and computational complexity.

In the following, we first present the network training performance of Deep-NOMA. Then we display some design examples of DeepMAS derived by end-to-end training. Finally, we provide the link-level simulation results of DeepNOMA and conventional schemes.

6.6.1 Network Training Performance

In this part, we respectively demonstrate the effectiveness of the multi-task balancing technique and ICNN-based inter-task interaction in training DeepNOMA.

6.6.1.1 Effect of Multi-task Balancing

We deploy multi-task balancing in end-to-end training of DeepNOMA under AWGN channel model in both OMA and NOMA cases, with $K = 4$ REs and $N = 4$ users for OMA, and $N = 6$ for NOMA. The size of message set is assumed as $M = 4$. Here we adopt the basic structure of DeepNOMA as proposed in Sect. 6.3 and use FC-DNN to parameterize all DNN blocks. Specifically, we apply 3 hidden layers with 16 neurons per layer in DeepMAS, and 4 hidden layers with 32 neurons per layer in DeepMUD for each task. A synthetic dataset is generated by randomly and uniformly drawing one-hot samples according to (6.16), and the size T of the dataset is empirically set as 40960. For each training epoch, the dataset is randomly shuffled, divided into small batches, and fed into DeepNOMA for end-to-end training. The training SNR is assumed as 1.5 dB and the balancing weight coefficient is empirically set as $b = 1$. For fairness, we assume equal weights among all tasks, i.e., $w_n = 1, \forall n$.

Figure 6.7 compares the training losses of DeepNOMA with or without the proposed multi-task balancing technique under the aforementioned settings. We use blue or green curves to respectively represent the training loss with or without task balancing, in short TB. The solid and dashed curves represent the average CE loss and the worst-user CE loss, respectively. The gray or blue-shaded area reflects the variation area of the CE loss of different tasks, respectively. For the OMA case, all users achieve the same performance whether there is balancing or not, as depicted in Fig. 6.7a. This is because each user can be provided with interference-free resources for data transmission in OMA, which makes the multi-user transmission non-interfereable.

However, multi-task balancing becomes indispensable in the NOMA case. To illustrate, we consider a pure NOMA setting in Fig. 6.7b with $N/K = 150\%$ overloading. If no balancing is assumed, all users compete for limited resources and some may outperform the others during training process, which causes a large gap between average CE loss and worst-user CE loss. After performing balancing, not only the fairness among multiple users is achieved, but also the overall loss is enhanced. The gains on average CE loss and worst-user CE loss also lead to significantly enhanced detection accuracies, as shown in the subplot of Fig. 6.7b. Therefore, multi-task balancing is crucial to enable fair and efficient training of DeepNOMA model in non-orthogonal mode, since the competitions among multiple correlated tasks may lead to great unfairness. In the following experiments, multi-task balancing is always adopted with the properly selected coeffient b.

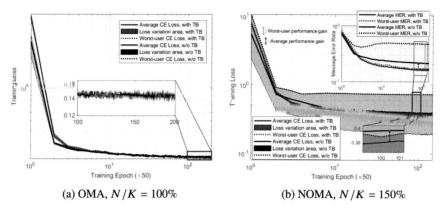

(a) OMA, $N/K = 100\%$ (b) NOMA, $N/K = 150\%$

Fig. 6.7 Comparisons on the training losses of DeepNOMA with or without the proposed multi-task balancing technique (in short, TB): **a** OMA case, and **b** NOMA case. End-to-end training in AWGN

6.6.1.2   Effect of ICNN in DeepMUD

We now analyze the effect of ICNN in DeepMUD. In Fig. 6.8, we deploy Deep-NOMA in a simple two-user case with $N = 2$ and $K = 1$, and show the end-to-end training loss and average message error rate (MER), where MER is defined in (6.4). For DeepMAS, we reuse the same setting as in Fig. 6.7. For DeepMUD, we consider a ICNN-based network structure as proposed in Sect. 6.5. We use $L = 1$ IC stage in DeepMUD, where $g_n^{(1),1}(\cdot)$ is parameterized by a 3-layer FC-DNN with 32 neurons per layer, and g_n^{Cls} and $g_n^{(1),\text{Cls}}$ are both parameterized by a 2-layer FC-DNN with 32 neurons per layer and a SoftMax layer as the output. The IC factor $\gamma_{n,j}^{(1)}$ is automatically learned during training. To show the performance of DeepMUD without IC, we simply fix $\gamma_{n,j}^{(1)} = 0, \forall n, j$, and train the network at 1 dB SNR. As shown in Fig. 6.8a, b, FC-DNN-based DeepMUD is enough to learn the optimal MUD with AWGN channel. Introducing ICNN does not enhance either training loss or MER. This is because the received signal in AWGN case is very simple. However, introducing ICNN in DeepMUD greatly improves the training loss and MER performances in Rayleigh fading channel, as illustrated in Fig. 6.8c, d. This is because fading channels greatly distort the input signal space, which cannot be easily learned by a simple DNN. Introducing IC in DeepMUD exploits the superimposed structure of input signal space and make it learnable with a practical network size. Therefore, ICNN is indispensable in DeepMUD to achieve high detection accuracy with fading channels.

In Fig. 6.9, we consider a typical code-domain NOMA case [14], with $N = 6$, $K = 4$ and $M = 4$, and analyze the training performance as well as the behavior of IC factor. Here we adopt a network setting of DeepNOMA similar to the one used in Fig. 6.8. Due to the increase of the user number, we enlarge the width of the layers in DeepMUD, where widths of the first and the second largest layers are set as 64 and 32

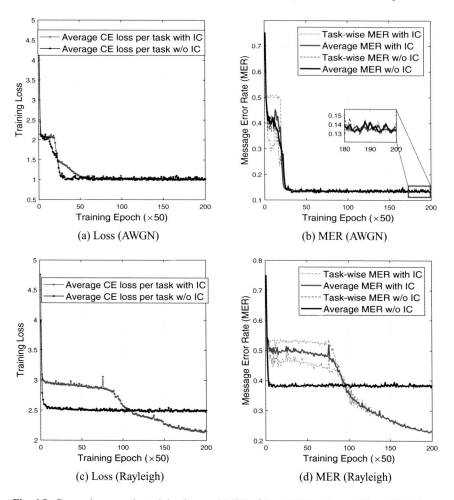

Fig. 6.8 Comparisons on the training loss and MER of DeepMUD with or without IC in Deep-NOMA: **a** AWGN channel, and **b** Rayleigh fading channel. End-to-end training of a simple two-user NOMA case with $N = 2$ and $K = 1$ is assumed

per task, respectively. The training is conducted under 1 dB SNR. Figure 6.9a depicts the losses during training, including the total loss $\mathcal{L}^{\text{Fading,Tot}}$, the task-wise CE loss $\mathcal{L}_n^{\text{CE}}$, task balancing loss $\mathcal{L}^{\text{B}}$, and local CE loss in IC stage $\mathcal{L}_n^{\text{CE-1}}$. Compared with the task-wise CE loss without IC, as illustrated by black-dotted curves, introducing IC can greatly enhance the classification performance. The MER performance is further shown in Fig. 6.9b, where introducing IC in DeepMUD can achieve a significant gain. The average value of IC factor, i.e., $\gamma = \mathbb{E}_{n,j}[\gamma_{n,j}^{(l)}]$, during training is presented in Fig. 6.9c. It is observed that γ converges to 1 for different system settings and random initializations. This shows that all inferred information shall be canceled during propagation to achieve better accuracy. DeepNOMA can also support larger

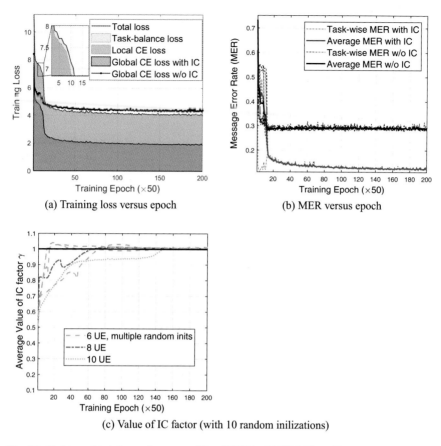

(a) Training loss versus epoch

(b) MER versus epoch

(c) Value of IC factor (with 10 random inilizations)

Fig. 6.9 End-to-end training performance of DeepNOMA with ICNN-based DeepMUD: **a** Training loss; **b** MER; **c** value of IC factor. Rayleigh fading with $K/N = 150\%$

NOMA systems with higher overloading. Figure 6.10 provides end-to-end training results with the overloading factors of 200% and 250%, where in both cases the training losses and MERs of DeepNOMA are reduced during training.

6.6.2   Design Examples of DeepMAS

In this part, we propose some design examples of DeepMAS. As a typical setting, we set $N = 6$, $K = 4$ and $M = 4$, and display the output of DeepMAS in both linear and non-linear modes. To generate linear DeepMAS, we assume QPSK modulation and parameterize the linear spreading signatures with learnable parameters. After end-to-end training, a learned signature set is shown in Fig. 6.11, where the l-th column is the normalized spreading signature of user-l.

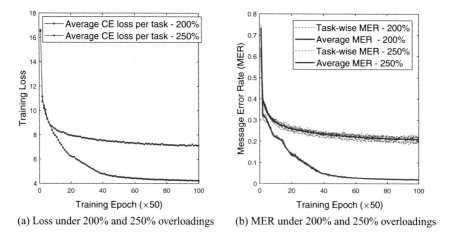

(a) Loss under 200% and 250% overloadings   (b) MER under 200% and 250% overloadings

Fig. 6.10 Performance of ICNN-based DeepNOMA for Rayleigh fading channel with various overloading factors

$$
S_{4\times6} = \begin{bmatrix}
0.53 + 0.035i & -0.39 - 0.32i & 0.16 - 0.25i & -0.11 - 0.46i & -0.25 + 0.38i & 0.50 - 0.43i \\
0.0043 - 0.10i & -0.33 + 0.40i & 0.32 + 0.42i & 0.22 - 0.24i & -0.76 + 0.0040i & -0.51 - 0.060i \\
0.35 - 0.23i & -0.0980 + 0.32i & 0.54 - 0.36i & 0.75 + 0.25i & 0.37 - 0.16i & 0.070 - 0.057i \\
0.62 + 0.37i & -0.33 + 0.50i & -0.44 - 0.10i & -0.15 - 0.16i & 0.07 - 0.23i & 0.0026 + 0.56i
\end{bmatrix}
$$
(6.48)

Fig. 6.11 A design example of linear DeepMAS

To generate non-linear DeepMAS, we set the constellation prior as parallelogram and conduct end-to-end training. Here all users are assumed to have the same total transmit power and channel model. Figure 6.12 depicts the multi-dimensional constellations on 4 REs learned by the 6 users, where different shapes represent different users. For illustration, we focus on the user represented by the green squares, and use the arrows to reflect the transmit powers of the complex symbols on different REs. In this case, the user has the minimum transmit power on RE-1, but has the maximum transmit power on RE-2 compared with the other users. Other users also hold diversified transmit powers on different REs, which naturally leads to the power gain differences among users. This fact promotes the efficiency of interference cancellation among users [21] and thus enhances the detection accuracy.

6.6.3 Performance Evaluation of DeepNOMA

In the following, we conduct link-level simulations to evaluate the performance of DeepNOMA in various channel models. In this part, we deploy 3 hidden layers with 16 neurons per layer in DeepMAS to learn non-linear MA signature mappings, and reuse the setting of DeepMUD in Fig. 6.9. The computational complexity is also analyzed.

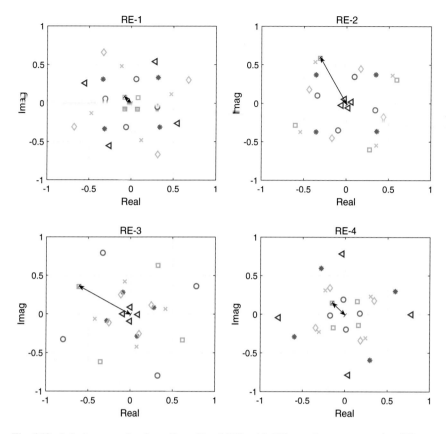

Fig. 6.12 A design example of non-linear DeepMAS, with different shapes representing different users. Power gain differences are learned on each RE

6.6.3.1 AWGN Channel

We perform end-to-end (E2E) training of DeepNOMA in AWGN channel at 7 dB with the setting of $N = 6$, $K = 4$ and $M = 4$. After the convergence of training, we acquire the trained DeepMAS and DeepMUD modules, and then compare the BER performance between the proposed scheme and conventional schemes. As illustrated in Fig. 6.13a, we consider typical MAS mappings including WBE [16], MUSA [13] and SCMA [21], and typical multi-user detectors including match-filter (MF), MMSE-SIC, and MPA with 5 inner iterations. It is seen that SCMA with MPA outperforms other conventional NOMA schemes, due to the gain provided by multi-dimensional constellation design as well as near-optimal MPA detector. Meanwhile, we observe that DeepNOMA achieves significant gains over SCMA, since Deep-NOMA is optimized in an end-to-end fashion while the transceivers of SCMA are designed separately.

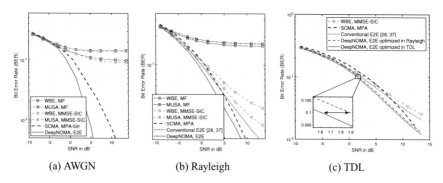

(a) AWGN   (b) Rayleigh   (c) TDL

Fig. 6.13 Performance of DeepMUD in fading channel models. Overloading factor $N/K = 150\%$

6.6.3.2 Fading Channels

We conduct an end-to-end training of DeepNOMA in fading channels with the same settings as in the AWGN case. Figure 6.13b shows the BER performance in Rayleigh block fading channel. Different from the AWGN case, fading channels leverage the performance gaps among conventional NOMA schemes. It is still observed that DeepNOMA achieves the lowest BER among all schemes, owing to the end-to-end training in the fading channel.

For practical implementation, DeepNOMA should be optimized offline in simple channel model and deployed online in realistic channel model. To validate the robustness of DeepNOMA against various channel models, we train it in Rayleigh fading channel and test it in tapped-delay-line (TDL) fading channel [12], where distance-dependent pathloss is considered. As shown in Fig. 6.13c, the specific DeepNOMA optimized in Rayleigh channel outperforms the conventional NOMA schemes. Fine-tuning the network parameters in TDL channel further provides a small bonus with respect to BER. These validate that, firstly, DeepNOMA has good generalization ability; and secondly DeepNOMA can be fine-tuned to the specific structure of channel models.

In Figs. 6.13b, c, we also provide the BER performances of the conventional end-to-end DL method [28, 37], where FC-DNN is used for joint detection. The DNN receiver used in the conventional method has 6 hidden layers, and the widths of the first and second largest layers are set as 256 and 128, which is much larger than DeepMUD. Nonetheless, DeepMUD still achieves significantly higher transmission accuracy, thanks to the sophisticated network design based on the domain expertise.

The computational complexity of the proposed DeepMUD and conventional methods is compared in Table 6.1. The complexity of conventional MUDs is given in [46], where N_{itr} is defined as the iteration number in the inner iteration of MPA and d_f is the largest number of superimposed user on all REs. For DeepMUD, we focus on the computational complexity involved in online deployment. Since input and output layers are rather narrow compared to the hidden layers, we can make a simplification on the complexity analysis by only considering the layers with largest

Table 6.1 Complexity analysis

Receiver	Complexity order	Normalized computation time in MATLAB
MPA	$O(N_{\mathrm{itr}} K d_f{}^2 M^{d_f})$	1
MMSE-SIC	$O(N K^3)$	0.67
DeepMUD	$O(S_1 S_2)$	0.32

widths. Therefore, DeepMUD approximately has a complexity order of $O(S_1 S_2)$, where S_1 and S_2 are the widths of the two consecutive layers with the largest widths. Besides, we count the computation time of various MUDs deployed in Fig. 6.13 in MATLAB. As shown in Table 6.1, DeepMUD can also reduce the computation time during online deployment. This benefit is due to the parallel detection among multiple tasks enabled by matrix multiplication in DNN, which facilitates the fast NOMA detection. Besides, DeepMUD can adapt to various NOMA schemes so long as the scheme-specific parameter set is applied, while MPA and MMSE-SIC can only deal with non-linear and linear spreading-based NOMA schemes, respectively.

To conclude, DeepNOMA simultaneously achieves better transmission accuracy and lower computational delay compared to conventional NOMA schemes. Deep-NOMA is also a universal framework for transceiver optimization of NOMA.

6.7 Conclusions

In this chapter, we proposed a unified framework for NOMA based on deep multi-task learning, namely DeepNOMA, which provides new insights into a principled route to the improvement of NOMA transceivers. The core idea behind our approach was to apply deep multi-task learning framework by treating non-orthogonal signal transmissions as multiple distinctive but interrelated tasks. A novel multi-task balancing loss function was then proposed to ensure fairness among tasks and to avoid stuck into local optima, which were validated with elaborate experiments.

Our proposed DeepNOMA was composed of DeepMAS and DeepMUD, corresponding to signature mapping and multi-user detection, respectively. Communication domain expert knowledge were exploited in the network structure design. Specifically, DeepMAS integrated the rules of desired alphabet and thus could generate signatures with regular shapes. Meanwhile, we exploited the superposition nature of the received signals and introduced the IC-based inter-task connection networks into DeepMUD, which enabled computational-efficient detection in fading channels. We trained DeepNOMA on synthetic dataset, and then performed detailed link level simulations to show the advantages of DeepNOMA over conventional schemes with respect to both BER and computational complexity. Due to the universal approximation property of DNN, DeepNOMA was also a unified framework for linear or non-linear, power or code domain, and grant-based or grant-free NOMA [28]. As a future direction, we will extend the proposed framework into multiple-antenna case. Parameter sharing or recurrent network structure may be exploited to reduce the network complexity in this case. Meanwhile, further theoretical analysis is also needed for better understanding the effect of incorporating DL in NOMA.

Appendix 1 Proof of Theorem 6.1

While the proposed framework works with general weight coefficients, this proof assumes that all tasks have equal weights as in [12–14], i.e., $w_n = 1, \forall n$, for simplicity of notations. With $w_n = 1$, (6.8) can further be simplified as

$$
\begin{aligned}
&\mathcal{L}\left([f_i]_{i=1}^N , [g_i]_{i=1}^N \right) \\
&= \mathbb{E}_{P(\mathcal{M})}\mathbb{E}_{P(\mathbf{y}|\mathcal{M};[f_i]_{i=1}^N)} \left[\sum_{n=1}^N \log P(\mathcal{M}_n|\mathbf{y}; g_n) \right] \\
&= \mathbb{E}_{P(\mathcal{M})}\mathbb{E}_{P(\mathbf{y}|\mathcal{M};[f_i]_{i=1}^N)} \left[\log \prod_{n=1}^N P(\mathcal{M}_n|\mathbf{y}; g_n) \right] \\
&= \mathbb{E}_{P(\mathcal{M})} \left[\mathbb{E}_{P(\mathbf{y}|\mathcal{M};[f_i]_{i=1}^N)} \left[\log P(\mathcal{M}|\mathbf{y}; [g_i]_{i=1}^N) \right] \right],
\end{aligned}
\tag{6.49}
$$

where $\prod_{n=1}^N P(\mathcal{M}_n|\mathbf{y}; g_n)$ is written as $P(\mathcal{M}|\mathbf{y}; [g_i]_{i=1}^N)$, since the estimated message $\mathcal{M}_n, \forall n$, only depends on g_n and is independent with each other.

Assume $\left[f_i^* \right]_{i=1}^N$ and $\left[g_i^* \right]_{i=1}^N$ as the optimal NOMA transceivers which maximize $\mathcal{P}1$. According to the definitions in (6.3), the output space of g_n^* has finite number of points. Hence, g_n^* is Lebesgue-integrable functions and is thus measurable. According to the Corollary 2.1 of [41], there exists a DNN that approximates any measurable function to any desired degree of accuracy. Therefore, we can find a set of $\left[g_i^{DNN} \right]_{i=1}^N$ which can arbitrarily approximate $\left[g_i^* \right]_{i=1}^N$, such that

$$
\sup_{\mathcal{M},\mathbf{y}} \left| \log P(\mathcal{M}|\mathbf{y}; \left[g_i^* \right]_{i=1}^N) - \log P(\mathcal{M}|\mathbf{y}; \left[g_i^{DNN} \right]_{i=1}^N) \right| < \epsilon,
\tag{6.50}
$$

where $\mathcal{M}$ has a finite support. Furthermore, we observe that f_n^* in (6.1) has finite support, since f_n^* takes $\mathcal{M}$ as input. According to Corollary 2.5 of [41], exact representation of f_n^* is possible with a DNN, denoted as f_i^{DNN}, i.e.,

$$
\left| P(\mathbf{y}|\mathcal{M}; \left[f_i^* \right]_{i=1}^N) - P(\mathbf{y}|\mathcal{M}; \left[f_i^{DNN} \right]_{i=1}^N) \right| = 0, \forall \mathcal{M}.
\tag{6.51}
$$

Based on (6.51) and (6.50), we have

$$\mathcal{A} = \sup_{\mathcal{M}} \left| \mathbb{E}_{P(\mathbf{y}|\mathcal{M};[f_i^*]_{i=1}^N)} \left[\log P(\mathcal{M}|\mathbf{y}; [g_i^*]_{i=1}^N) \right] \right.$$
$$\left. - \mathbb{E}_{P(\mathbf{y}|\mathcal{M};[f_i^{\mathrm{DNN}}]_{i=1}^N)} \left[\log P(\mathcal{M}|\mathbf{y}; [g_i^{\mathrm{DNN}}]_{i=1}^N) \right] \right|$$
$$\leq \sup_{\mathcal{M}} \left\{ \mathbb{E}_{P(\mathbf{y}|\mathcal{M};[f_i^*]_{i=1}^N)} \left[\sup_{\mathcal{M},\mathbf{y}} \left| \log P(\mathcal{M}|\mathbf{y}; [g_i^*]_{i=1}^N) \right. \right. \right.$$
$$\left. \left. \left. - \log P(\mathcal{M}|\mathbf{y}; [g_i^{\mathrm{DNN}}]_{i=1}^N) \right| \right] \right\} < \sup_{\mathcal{M}} \{\epsilon\} = \epsilon. \tag{6.52}$$

The left hand side (LHS) of (6.9) is then constrained by

$$\text{LHS of (6.9)} \leq \mathbb{E}_{P(\mathcal{M})}[\mathcal{A}] < \mathbb{E}_{P(\mathcal{M})}[\epsilon] = \epsilon. \tag{6.53}$$

Appendix 2 Proof of Theorem 6.2

The K-dimensional complex received vector can be equivalently represented by a $2K$-dimensional real vector $\mathbf{y}$ [20], which follows

$$P(\mathbf{y}) = \sum_{\substack{\mathbf{x}_n \in \mathcal{X}_n, \\ n=1 \cdots N}} \frac{1}{\prod\limits_{s=1}^N |\mathcal{X}_s|} \mathcal{N}\left(\sum_{n=1}^N [\mathrm{Re}(\mathbf{x}_n) \ \mathrm{Im}(\mathbf{x}_n)], \sigma_0^2 \mathbf{I} \right).$$

We assume that $\mathbf{y}$ has zero mean and set $q = 2K$ for brevity. Define $\mathbf{y}_N$ as a q-dimensional normal-distributed random vector with the same mean and co-variance as $\mathbf{y}$. Then we have

$$h(\mathbf{y}) \leq h(\mathbf{y}_N) = \frac{1}{2} \log \left[(2\pi e)^q \left| (\Sigma_{\mathbf{y}}) \right| \right] = h^U(\mathbf{y}), \tag{6.54}$$

according to maximum differential entropy lemma. We rewrite NG$(\mathbf{y})$ as

$$\mathrm{NG}(\mathbf{y}) = \mathrm{CE}\left(P(\mathbf{y}), P(\mathbf{y}_N) \right) - h(\mathbf{y}), \tag{6.55}$$

where $\mathrm{CE}\left(P(\mathbf{y}), P(\mathbf{y}_N) \right)$ represents the cross-entropy function, and is simplified as

$$\mathrm{CE}(P(\mathbf{y}), P(\mathbf{y}_N)) = -\int P(\mathbf{y}) \log \left(\mathcal{N}\left(\mu_{\mathbf{y}}, \Sigma_{\mathbf{y}} \right) \right) d\mathbf{y}$$
$$= -\mathbb{E}_{P(\mathbf{y})} \left[-\frac{1}{2} \mathbf{y}^T \Sigma_{\mathbf{y}}^{-1} \mathbf{y} - \frac{1}{2} \log \left((2\pi e)^q |\Sigma_{\mathbf{y}}| \right) \right]$$
$$= \frac{1}{2} \left(\log \left((2\pi e)^q |\Sigma_{\mathbf{y}}| \right) + \mathbb{E}_{P(\mathbf{y})} \left[\mathbf{y}^T \Sigma_{\mathbf{y}}^{-1} \mathbf{y} \right] \right)$$
$$= \frac{1}{2} \left(\log \left((2\pi e)^q |\Sigma_{\mathbf{y}}| \right) + \mathrm{tr}[\mathbf{I}] \right) = h^U(\mathbf{y}) + \text{constant}. \tag{6.56}$$

Aggregating (6.55) and (6.56), we have

$$h(\mathbf{y}) = -\left(\mathrm{NG}\,(\mathbf{y}) - h^U\,(\mathbf{y})\right),\tag{6.57}$$

up to a constant. Therefore, maximizing $h(\mathbf{y})$ in $\mathcal{P}4$ is equivalent to minimizing $\mathrm{NG}\,(\mathbf{y}) - h^U\,(\mathbf{y})$. Furthermore, since $h^U\,(\mathbf{Y})$ is determined by the co-variance of $\mathbf{y}$, this value represents the power of the composite signal, which is approximately constant in the case of NOMA.

Appendix 3 Proof of Theorem 6.3

We prove this theorem by first constructing a DNN block to recast the joint typicality decoding [10] and then bounding the width of this DNN.

The principle of joint typicality decoding is to find the unique message such that this message and the received signal are jointly typical [10]. Denote $(x^n, y^n) \in (\mathcal{X}^n, \mathcal{Y}^n)$ as a pair of sequences with length-n. Let $(X, Y) \sim p(x, y)$. Without loss of generality, we assume $|\mathcal{Y}| = M$. Define $\mathcal{T}_\epsilon^{(n)}(X)$ and $\mathcal{T}_\epsilon^{(n)}(X, Y)$ as the sets of ϵ-typical and jointly ϵ-typical sequence pair with length-n, respectively.

With joint typicality decoding, the decoding error probability is bounded by the conditional typicality lemma, which states that, given the transmitted signal sequence $x^n \in \mathcal{T}_{\epsilon'}^{(n)}(X)$ and the received signal sequence $Y^n \sim \prod_{i=1}^n p_{Y|X}(y_i|x_i)$, the following limit holds for any $\epsilon > \epsilon'$ with sufficiently large n

$$P\{(x^n, Y^n) \in \mathcal{T}_\epsilon^{(n)}(X, Y)\} \to 1.\tag{6.58}$$

Now we propose a DNN block which performs joint typicality decoding during forward-propagation, as illustrated in Fig. 6.14. In this DNN, Layer-1 calculates the frequency of each elements of $\mathcal{Y}$. Layer-2 compares the empirical PMF of y^n with its conditional PDF given $x_i^n \in \mathcal{T}_{\epsilon'}^{(n)}(X)$. The output layer demonstrates whether y^n and x_i^n are jointly typical by activating or deactivating the corresponding neuron. Since $|\mathcal{T}_\epsilon^{(n)}(X)| \doteq 2^{nH(X)}$, this DNN is formed by $2^{nH(X)}$ parallel branches, with the i-th branch corresponding to a unique message x_i^n, $1 \leq i \leq 2^{nH(X)}$.

We focus on the i-th branch. Layer-1 in the i-th branch contains M sub-networks each with width-n. Given y^n, the output $\boldsymbol{\alpha}_{i,m}^{(1)}$ of the m-th sub-network, $1 \leq m \leq M$, in the i-th branch is given by

$$\boldsymbol{\alpha}_{i,m}^{(1)} = \sigma_h\left(\mathbf{I}y^n - y_m\mathbf{1}\right),\tag{6.59}$$

where $\mathbf{I}$ is identity matrix, $\mathbf{1}$ is the all 1's vector with length-n, and $y_m \in \mathcal{Y}$ is the m-th alphabet in $\mathcal{Y}$. The activation function is defined as $\sigma_h(t) = \begin{cases} 0, & t \leq -h \text{ or } t \geq h, \\ 1, & -h < t < h, \end{cases}$
where h is a relative small value and follows $h < \min_{y_j, y_m \in \mathcal{Y}} |y_j - y_m|$.

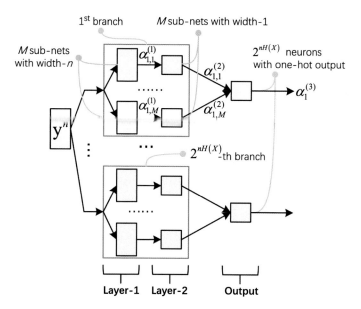

Fig. 6.14 The jointly typical decoding based on DNN

Layer-2 in the i-th branch also contains M sub-networks each with width-1. The activated value $\alpha_{i,m}^{(2)}$ of the m-th sub-network of layer-2 in the i-th branch is given by

$$\alpha_{i,m}^{(2)} = \sigma_\epsilon \left(\frac{1}{n} \mathbf{1}^\top \alpha_{i,m}^{(1)} - P(y_m|x_i^n) \right), \tag{6.60}$$

where $x_i^n \in \mathcal{T}_{\epsilon'}^{(n)}(X)$ is the i-th branch typical sequence, and $P(y_m|x_i^n)$ is the conditional probability of y_m given x_i^n. The activation function is defined as $\sigma_\epsilon(t) =$
$$\begin{cases} 1, & t \leq -\epsilon \text{ or } t \geq \epsilon, \\ 0, & -\epsilon < t < \epsilon. \end{cases}$$
The output layer corresponding to the i-th branch performs an add operation, i.e.,

$$\alpha_i^{(3)} = \mathbf{1}\left[\alpha_{i,1}^{(2)}, \cdots, \alpha_{i,M}^{(2)}\right]. \tag{6.61}$$

Due to the conditional typicality lemma, only one neuron in the output layer will be activated, whose index corresponds to the estimated sequence with the error probability bound by ϵ. Since Layer-1 has the largest width, the width of the proposed DNN in Fig. 6.14 is bounded by

$$nM \times 2^{nH(X)} \overset{(a)}{\leq} nM \times M^n = O(nM^n), \tag{6.62}$$

where (a) is due to the maximum entropy principle.

Now we consider a NOMA system with the transmission signal pair (X_1, X_2) and the received signal Y. Using ICNN structure in Fig. 6.5, the joint typicality of (X_1, Y) and (X_2, Y) shall be examined at stage-1 and stage-2 of ICNN, respectively. For each stage, the DNN proposed in Fig. 6.14 would be enough to detect source message. Therefore, the width is bounded by $O(nM^n)$. On the other hand, if we use FC-DNN which recasts the multivariate joint typical decoding, the network requires $2^{nH(X_1)}2^{nH(X_2)}$ branches, leading to a width bound of $O(nM^{2n})$.

References

1. K. Yang et al., Non-orthogonal multiple access: achieving sustainable future radio access. IEEE Commun. Mag. **57**(2), 116–121 (2019)
2. Z. Ding et al., Application of non-orthogonal multiple access in LTE and 5G networks. IEEE Commun. Mag. **55**(2), 185–191 (2017)
3. Y. LeCun, Y. Bengio, G. Hinton, Deep learning. Nature **521**(7553), 436–444 (2015)
4. T. Wang et al., Deep learning for wireless physical layer: opportunities and challenges. China Commun. **14**(11), 92–111 (2017)
5. Z. Qin, H. Ye, G.Y. Li, B.F. Juang, Deep learning in physical layer communications. IEEE Wirel. Commun. **26**(2), 93–99 (2019)
6. H. He et al., Model-driven deep learning for physical layer communications. IEEE Wirel. Commun. **26**(5), 77–83 (2019)
7. B. Zhu, J. Wang, L. He, J. Song, Joint transceiver optimization for wireless communication PHY using neural network. IEEE J. Sel. Areas Commun. **37**(6), 1364–1373 (2019)
8. X. You et al., AI for 5G: research directions and paradigms. Sci. China Inf. Sci. **62**(2), 21301 (2018)
9. T. O'Shea, J. Hoydis, An introduction to deep learning for the physical layer. IEEE Trans. Cogn. Commun. Netw. **3**(4), 563–575 (2017)
10. A.E. Gamal, Y.-H. Kim, *Network Information Theory* (Cambridge Univ. Press, Cambridge, U.K., 2011)
11. M. Vaezi, R. Schober, Z. Ding, H.V. Poor, Non-orthogonal multiple access: common myths and critical questions (2018). arxiv:1809.07224
12. 3GPP TR 38.812, Study on non-orthogonal multiple access (NOMA) for NR (Release 15). 3rd generation partnership project; technical specification group radio access network (2018)
13. Z. Yuan et al., Multi-user shared access for Internet of Things, in *Proceedings of the IEEE 83rd VTC Spring* (IEEE, Nanjing, 2016), pp. 1–5
14. H. Nikopour, H. Baligh, Sparse code multiple access, in *Proceedings of the IEEE 24th PIMRC* (IEEE, London, 2013), pp. 332–336
15. X. Dai, Z. Zhang, B. Bai, S. Chen, S. Sun, Pattern division multiple access: a new multiple access technology for 5G. IEEE Wirel. Commun. **25**(2), 54–60 (2018)
16. W. Liu, X. Hou, L. Chen, Enhanced uplink non-orthogonal multiple access for 5G and beyond systems. Front. Inf. Technol. Electron. Eng. **19**(3), 340–356 (2018)
17. 3GPP, R1-1808499, Transmitter side signal processing schemes for NCMA, LG Electronics. RAN1#94
18. L. Yu, P. Fan, D. Cai, Z. Ma, Design and analysis of SCMA codebook based on Star-QAM signaling constellations. IEEE Trans. Veh. Technol. **67**(11), 10543–10553 (2018)
19. K. Xiao et al., On capacity-based codebook design and advanced decoding for sparse code multiple access Systems. IEEE Trans. Wirel. Commun. **17**(6), 3834–3849 (2018)
20. N. Ye et al., On constellation rotation of NOMA with SIC receiver. IEEE Commun. Lett. **22**(3), 514–517 (2018)

21. M. Taherzadeh, H. Nikopour, A. Bayesteh, H. Baligh, SCMA codebook design, in *Proceedings of the 2014 IEEE 80th Vehicular Technology Conference (VTC2014-Fall)* (IEEE, Vancouver, BC, 2014), pp. 1–5

22. L. Dai et al., A survey of non-orthogonal multiple access for 5G. IEEE Commun. Surv. Tutor. **20**(3), 2294–2323 (2018)

23. F. Wei, W. Chen, Low complexity iterative receiver design for sparse code multiple access. IEEE Trans. Commun. **65**(2), 621–634 (2017)

24. J. Dai, K. Niu, C. Dong, J. Lin, Improved message passing algorithms for sparse code multiple access. IEEE Trans. Veh. Technol. **66**(11), 9986–9999 (2017)

25. L. Yuan, J. Pan, N. Yang, Z. Ding, J. Yuan, Successive interference cancellation for LDPC-coded non-orthogonal multiple access systems. IEEE Trans. Veh. Technol. **67**(6), 5460–5464 (2018)

26. B.K. Jeong, B. Shim, K.B. Lee, MAP-based active user and data detection for massive machine-type communications. IEEE Trans. Veh. Technol. **67**(9), 8481–8494 (2018)

27. Q. Wang, R. Zhang, L. Yang, L. Hanzo, Non-orthogonal multiple access: a unified perspective. IEEE Wirel. Commun. **25**(2), 10–16 (2018)

28. N. Ye, X. Li, H. Yu, A. Wang, W. Liu, X. Hou, Deep learning aided grant-free NOMA towards reliable low-latency access in tactile Internet of Things. IEEE Trans. Ind. Inform. **15**(5), 2995–3005 (2019)

29. E. Nachmani et al., Deep learning methods for improved decoding of linear codes. IEEE J. Sel. Top. Signal Process. **12**(1), 119–131 (2018)

30. H. Ye, G.Y. Li, Initial results on deep learning for joint channel equalization and decoding, in *Proceedings of the 2017 IEEE 86th Vehicular Technology Conference (VTC-Fall)* (IEEE, Toronto, ON, 2017), pp. 1–5

31. H. Huang, J. Yang, H. Huang, Y. Song, G. Gui, Deep learning for super-resolution channel estimation and DOA estimation based massive MIMO system. IEEE Trans. Veh. Technol. **67**(9), 8549–8560 (2018)

32. J. Xu, P. Zhu, J. Li, X. You, Deep learning based pilot design for multi-user distributed massive MIMO systems. IEEE Wirel. Commun. Lett. **8**(4), 1016–1019 (2019)

33. T.J. O'Shea, T. Erpek, T.C. Clancy, Physical layer deep learning of encodings for the MIMO fading channel, in *Proceedings of the 55th Annual Allerton Conference on Communication, Control, and Computing* (Monticello, IL, 2017), pp. 76–80

34. T.J. O'Shea, T. Roy, N. West, B.C. Hilburn, Physical layer communications system design over-the-air using adversarial networks, in *Proceedings of the 26th EUSIPCO* (Rome, 2018), pp. 529–532

35. A. Anderson, S.R. Young, T.P. Karnowski, J.M. Vann, Deepmod: an over-the-air trainable machine modem for resilient PHY layer communications, in *Proceedings of the IEEE MILCOM* (Los Angeles, CA, USA, 2018), pp. 213–218

36. F. Sun, K. Niu, C. Dong, Deep learning based joint detection and decoding of non-orthogonal multiple access systems, in *Proceedings of the IEEE Globecom WS* (Abu Dhabi, United Arab Emirates, 2018), pp. 1–5

37. M. Kim et al., Deep learning-aided SCMA. IEEE Commun. Lett. **22**(4), 720–723 (2018)

38. G. Gui, H. Huang, Y. Song, H. Sari, Deep learning for an effective non-orthogonal multiple access scheme. IEEE Trans. Veh. Technol. **67**(9), 8440–8450 (2018)

39. M. Liu, T. Song, G. Gui, Deep cognitive perspective: resource allocation for NOMA based heterogeneous IoT with imperfect SIC. IEEE Internet Things J. **6**(2), 2885–2894 (2019)

40. D.P. Kingma, M. Welling, Auto-encoding variational Bayes, in *Proceedings of the 2nd International Conference on Learning Representations* (Banff, Canada, 2014), pp. 1–14

41. K. Hornik, M. Stinchcombe, H. White, Multilayer feedforward networks are universal approximators. Neural Netw. **2**(5), 359–366 (1989)

42. A. Zappone, et al., Model-aided wireless artificial intelligence: embedding expert knowledge in deep neural networks towards wireless systems optimization (2018). arxiv:1808.01672

43. S. Ruder, An overview of multi-task learning in deep neural networks (2017). arxiv:1706.05098

44. H. Yu et al., Optimal design of resource element mapping for sparse spreading non-orthogonal multiple access. IEEE Wirel. Commun. Lett. **7**(5), 744–747 (2018)
45. X. Glorot, Y. Bengio, Understanding the difficulty of training deep feedforward neural networks, in *Proceedings of the AISTATS* (IEEE, 2010), pp. 249–256
46. 3GPP R1-165619, Transceiver implementation and complexity analysis for SCMA (Huawei, HiSilicon, 2016)

Chapter 7
Deep Learning-Aided High-Throughput Multiple Access

Abstract In this chapter, we discuss the deep learning-aided high-throughput multiple access. Section 7.1 introduces the motivation of applying deep learning to grant-free NOMA in tactile Internet of Things. Section 7.2 introduces the system model of grant-free NOMA. Section 7.3 presents the neural network model for grant-free NOMA. Section 7.4 analyzes the loss function and the training algorithm of the proposed network model. Section 7.5 presents the evaluation results to validate the feasibility and efficiency of the proposed scheme. Section 7.6 draws the conclusions.

7.1 Introduction

Internet of Things (IoT) has been regarded as a bright new business model envisioned by next generation wireless communication (5G) to connect and control tremendous number of real and virtual objects [1, 2]. One important variant of IoT is the mission critical IoT [3], which requires ultra-responsive and ultra-reliable connectivities between objects and networks [4]. The Tactile Internet [5], targeting at ultra-low latency communications together with high availability, reliability, and security, has been regarded as a promising enabler of mission critical IoT services [3]. Tactile IoT warrants a big paradigm change from the conventional data-delivery networks to technology-transfer networks, and accordingly reforms virtually almost all parts of the society [4, 5].

The researches of physical layer technologies in 5G have contributed to addressing the ultra-low latency challenge of Tactile IoT while ensuring certain level of reliability [6], such as short-block channel coding [7], waveform, numerology, and frame structure [4]. A recent outdoor experimental trail has validated that reliable point-to-point transmission can be ensured with near 1 ms round trip latency by employing 5G technologies [8]. However, existing researches towards Tactile IoT mainly concentrate on the point-to-point communication link, which is only able to connect a single haptic device. Nonetheless, to approach the next level of immersion, multi-model data from massive distributed or area-based IoT devices with sporadic traffic are simultaneously required [1, 4, 9]. This poses great challenge on the design of physical layer technologies due to the existence of multiple transmitters and limited radio resources.

To solve this challenge, grant-free non-orthogonal multiple access (NOMA), which exploits the joint benefit of grant-less access mechanism and non-orthogonal signal superposition, has been considered to simultaneously realize low latency and high-efficient massive access in Tactile IoT [10–12]. In grant-based transmission, a random access signaling interaction and a scheduling grant are required from the base station (BS) before data transmission, which normally costs tens of milliseconds. In grant-free NOMA, data transmissions are autonomously activated by the users without explicit dynamic grant, which greatly reduces the control/user plane latency caused by signaling interaction and scheduling grant [13]. However, due to the decentralized instant transmissions in grant-free NOMA, the activity information is unknown at the receiver. Moreover, the unexpected inter-user interference leads to the deteriorated transmission reliability, which dissatisfies the demand of Tactile IoT.

Grant-based NOMA has been designed to enhance the transmission reliability based on capacity optimization [14–18], following the classical multi-user information theory [19]. However, grant-free NOMA does not obey the conventional Shannon information theory [20], and is usually designed empirically [21–25]. In [21–23], grant-free mechanism is directly introduced into the state-of-art NOMA schemes, i.e., sparse code multiple access (SCMA), multi-user shared access (MUSA) and pattern division multiple access (PDMA), which are categorized according to their specially designed spreading signatures. Besides, in [24], grant-free NOMA is optimized for short packet transmission, while in [25], a multi-branch structure with unequal protection per branch is proposed to achieve enhanced reliability. As for receiver design, some resort to compressive sensing-based approaches [12, 26, 27], and some consider the message passing algorithm-based methods [28, 29]. Although the existing schemes have significantly contributed to the low latency multi-user radio access, few have focused on the end-to-end optimization of the entire grant-free NOMA system. The major obstacle lays in the fact that the introduction of random user activation behavior and non-orthogonal transmissions makes the grant-free NOMA non-trivial to be mathematically modeled and optimized by traditional optimization methods. To this end, this chapter focuses on the end-to-end optimization of grant-free NOMA based on deep learning (DL).

In the most recent decade, DL has achieved great success in solving very complicated optimization problems in a data-driven fashion [30]. For the supervised learning problem which aims to learn the mapping function between input data and corresponding label, DL mimics the target function with deep neural network (DNN), which acts as a global function approximator, and then optimizes the network parameters via, e.g., stochastic gradient decent (SGD), to minimize the function approximation error. For the unsupervised learning problem which aims to extract the inner features of unlabeled data, one means is employing deep auto-encoder (DAE) to reconstruct input data and regard the representations at the hidden layer as the inner features. Deep variational auto-encoder (VAE) is a recent proposed variant of DAE, which is built on the variational inference theory [31].

The booming of DL has shed a new light on optimizing the wireless communication system [32–34]. Nevertheless, existing DL models are usually designed for human-level tasks, and may not ideally suit wireless communications especially the grant-free NOMA. In some initial works, DL has been introduced to improve the NOMA systems [35, 36] with the mode of offline training and online operating. The authors in [35] introduce a DAE model for grant-based NOMA which achieves better accuracy than the conventional SCMA. Gui et al. [36] exploits the long-term dependence of NOMA system to train the input signals by incorporating long short term memory cells. However, these methods focus on the grant based cases, and do not consider the random user activation, which should be modeled in grant-free NOMA.

As above mentioned, existing grant-free NOMA schemes normally reuse the spreading signatures designed for grant-based NOMA. Besides, plenty of sophisticated human-crafted works are required to design good transceivers, which contradicts the trend of automatic communication in industrial automation. Moreover, the performance of grant-free NOMA is mathematically intractable, which is hard to be optimized. To tackle these challenges, a DL-aided grant-free NOMA scheme is proposed towards reliable low-latency access in Tactile IoT by a sophisticated integration between grant-free NOMA with DL. Aiming at optimizing the transmission reliability of grant-free NOMA, we formulate a variational optimization problem which incorporates the random user activation, symbol spreading and multi-user signal detection. Since the problem is intractable, we propose a neural network model for grant-free NOMA to parameterize the intractable variational functions to exploit the universal approximation property of DNNs. We optimize the parameters of the proposed network in a data-driven fashion, and derive the spreading signatures as well as the multi-user detector, which outperforms conventional grant-free NOMA schemes under fair comparisons. The proposed scheme constitutes a universal framework to design good transmitter and receiver for grant-free NOMA.

The contributions are summarized as follows:

1. We establish an end-to-end neural network model of grant-free NOMA based on deep VAE to parametrize the proposed variational optimization problem. The proposed network model consists of a NOMA encoding network, which models the random user activation and symbol spreading, and a NOMA decoding network, which jointly estimates the user activity and the transmitted symbols. The spreading signatures and receiving algorithm can be obtained according to the training results of the NOMA encoding and decoding networks, respectively.

2. We propose a novel multi-loss function to train the proposed network, which includes a symbol reconstruction loss and a user activity detection loss. Specifically, the prior information of user's activation probability is exploited by introducing the confidence penalty in the loss function.

3. We derive both linear and non-linear spreading signatures by training the proposed network, which show remarkable gain on the transmission reliability than conventional grant-free NOMA schemes. These signatures can be automatically generated, which perfectly matches the highly automatic applications of Tactile IoT.

7.2 System Model and Problem Formulation

As illustrated in Fig. 7.1, this chapter considers a typical uplink grant-free NOMA system with L users and a BS, where each user may activate and transmit signals in each time interval with the probability α, $0 < \alpha \leq 1$. The activation probabilities can either be acquired according to the traffic models of the users, or be estimated by the BS according to the historical data. Suppose that user-l, $1 \leq l \leq L$, becomes active in the t-th time interval, the source bits of user-l are first transformed into a modulated symbol $x_l^{(t)} \in \mathbb{C}$, which may take value from the constellation alphabet $\mathcal{X} = \{\mathcal{X}_1, \ldots, \mathcal{X}_m, \ldots, \mathcal{X}_M\}$ with $|\mathcal{X}| = M$, e.g. M-QAM. We note that, practically, the neural network only deals with real values. To this end, we use the two-dimensional real column vector $x_l^{(t)} = [\text{Re}\{x_l^{(t)}\}, \text{Im}\{x_l^{(t)}\}] \in \mathbb{R}^2$ to represent the complex modulation symbol, and adopt the real-valued signal model throughout this chapter. Hence, each complex value can be received by two neurons, where one neuron takes the real part and the other takes the imaginary part.

$x_l^{(t)}$ is then spread to generate a symbol sequence $s_l^{(t)} \in \mathbb{R}^{2K}$ according to the spreading signature of user-l, which is later transmitted over K subcarriers. We assume that $s_l^{(t)}$ has unit transmit power, i.e., $\|s_l^{(t)}\| = 1$. For users which are inactive in the t-th time interval, the transmitted signal can be regarded as a zero vector. For brevity, we omit (t) in the rest of the paper. In a typical overloaded NOMA system, K is usually smaller than L, so that multiple signal streams may overlap together, which constitute the key attribute of NOMA. The received signal vector is given by

$$y = \sum_{l=1}^{L} \mathcal{I}(l) \sqrt{P_l} \, \text{diag} \, (h_l) \, s_l + n, \qquad (7.1)$$

where $P_l \in \mathbb{R}^+$ is the transmit power of user-l, and $\mathcal{I}(l) \in \{0, 1\}$ is the indicating function which takes value "1" when user-l is active, or "0" otherwise. Besides, $h_l = [h_{l,1}, h_{l,2}, \ldots, h_{l,K}] \in \mathbb{R}^{2K}$ is the channel coefficient vector of user-l over K subcarriers, and $n \sim \mathcal{N}(n|0, \sigma_0^2 I)$ denotes the additive white Gaussian noise

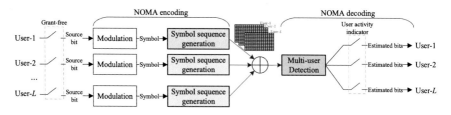

Fig. 7.1 System model of Grant-free NOMA. Physical resources are reserved for L users, and each user activates with a certain probability. The source bits of each user are first modulated and then mapped to a symbol sequence according to user-specific symbol spreading signature. The receiver jointly detects the user activity as well as source bits

(AWGN) vector with variance σ_0^2. In this chapter, we assume that users experience quasi-static channel environment where the channel coefficients remain unchanged for quite a long time.

Due to the grant-free access mechanism, the BS has to jointly identify the user activity and detect the transmitted signals. We define the estimated activity of user-l as $\hat{I}(l) \in \{0, 1\}$. The true positive (TP) event, where an active user is correctly identified by the BS, is defined as $\mathcal{E}_{TP} : \{I(l) = 1 \,\&\, \hat{I}(l) = 1\}$. And the false positive (FP) event, where an inactive user is mistakenly identified by the BS, is defined as $\mathcal{E}_{FP} : \{I(l) \neq 1 \,\&\, \hat{I}(l) = 1\}$. We denote $I = [I(1), \ldots, I(l), \ldots, I(L)]$, and $\hat{I} = [\hat{I}(1), \ldots, \hat{I}(l), \ldots, \hat{I}(L)]$. Similarly, we define the estimated modulated symbol for the active user-l as $\hat{x}_l$, and the symbol error event is defined as $\mathcal{E}_E : \{x_l \neq \hat{x}_l\}$.

To promote the reliability of the grant-free NOMA, the modulation symbol to sequence mapping should be elaborately designed to ease the signal separation at the multi-user detector, and the detector should also be properly optimized to enhance the accuracies of both user activity and modulation symbol detections. Despite these clear goals, it is, however, non-trivial to directly optimize the components of the grant-free NOMA system via conventional optimization techniques. The main difficulties lay in the intractability of the optimization problem, and the high computational complexity involved in searching good solutions. Therefore, we resort to DL to find practical and efficient method to optimize the grant-free NOMA, where DL has shown great capability in optimizing very complicated and even mathematically intractable problems in a data-driven fashion.

7.3 Deep Learning-Aided Grant-Free NOMA

In this section, we exploit the reciprocity between the deep VAE model and grant-free NOMA, and then propose a specified network architecture of DL-aided grant-free NOMA.

7.3.1 Deep VAE for Grant-Free NOMA

As a variant of deep auto-encoding neural networks, VAE is able to extract compact latent representations of the observed large amount of data using both variational inference and neural network [31]. VAE consists of an probabilistic encoding network and a probabilistic decoding network. One observed datapoint d is first fed to the probabilistic encoder to produce a distribution $q_\phi(z|d)$, over the possible values of the latent code z. Afterwards, one or several samples of the latent code are sampled from $q_\phi(z|d)$ and fed to the probabilistic decoder, denoted by $p_\theta(\hat{d}|z)$, which generates the distribution of the estimated datapoint $\hat{d}$. If we regard s as the generated modulation symbol and choose Gaussian distribution as the prior of $q_\phi(z|s)$, we observe that VAE achieves the similar goal as NOMA, where the encoder and the decoder networks in VAE correspond to NOMA encoding and decoding procedures, respectively. This motivates us to design grant-free NOMA via a variational Bayes auto-encoding perspective as follows.

Define $x = [x_1^T, \dots, x_l^T, \dots, x_L^T]^T \in \mathbb{R}^{2L}$, $x_l \in \{X, 0\}$, as the column vector of the transmitted symbol in grant-free NOMA. In NOMA, x should be mapped to a latent code, referred to as s, whose noise-polluted version $\tilde{s}$ best regenerates x. However, the optimal mapping between x and $\tilde{s}$ is generally unknown. Assume that there exists an optimal mapping, and we denote the joint distribution of x and $\tilde{s}$ under the optimal mapping relation as

$$p_{\theta^*}(\tilde{s}, x) = p_{\theta^*}(\tilde{s}|x)p(x), \tag{7.2}$$

where θ^* represents the model parameters, and $p(x)$ is determined only by α and X, as follows

$$p(x_l) = \begin{cases} 1 - \alpha, & x_l = 0, \\ \alpha/M, & x_l = X_m \in X, \end{cases} \tag{7.3}$$

Unfortunately, θ^* is normally unknown to us, and we thus resort to variational inference to approximate $p_{\theta^*}(\tilde{s}|x)$ with a family of distributions $q_\phi(\tilde{s}|x)$, where ϕ is the variational parameter. The approximation error can be measured by Kullback-Leibler divergence (KLD)

$$\begin{aligned} \mathrm{KL}(q_\phi(\tilde{s}|x)||p_{\theta^*}(\tilde{s}|x)) &= \mathbb{E}_{q_\phi(\tilde{s}|x)}\left[\log q_\phi(\tilde{s}|x)\right] \\ &\quad - \mathbb{E}_{q_\phi(\tilde{s}|x)}\left[\log p_{\theta^*}(\tilde{s}, x)\right] + \log p(x). \end{aligned} \tag{7.4}$$

To minimize the approximation error caused by $q_\phi(\tilde{s}|x)$, KLD in (7.4) should be minimized. Notice that the third right-hand side (RHS) term in (7.4) is a constant, and we readily see that minimizing (7.4) is equivalent to maximizing

$$\begin{aligned} \mathcal{L}(\theta^*, \phi|x) &= \mathbb{E}_{q_\phi(\tilde{s}|x)}\left[\log p_{\theta^*}(\tilde{s}, x)\right] - \mathbb{E}_{q_\phi(\tilde{s}|x)}\left[\log q_\phi(\tilde{s}|x)\right] \\ &= \mathbb{E}_{q_\phi(\tilde{s}|x)}\left[\log p_{\theta^*}(x|\tilde{s})p_{\theta^*}(\tilde{s})\right] - \mathbb{E}_{q_\phi(\tilde{s}|x)}\left[\log q_\phi(\tilde{s}|x)\right] \\ &= \mathbb{E}_{q_\phi(\tilde{s}|x)}\left[\log p_{\theta^*}(x|\tilde{s})\right] - \mathrm{KL}(q_\phi(\tilde{s}|x)||p_{\theta^*}(\tilde{s})). \end{aligned} \tag{7.5}$$

The first RHS term in (7.5) is the expected negative log-likelihood, i.e., the reconstruction loss, which encourages the best regeneration of the source symbols from the latent codes. The second RHS term is the KLD between the posterior distributions of the latent codes given x and the prior distribution of the latent codes. We note that, to approach the optimal rate in AWGN channel, the constellation alphabet should follow the Gaussian distribution. Hence, we set $p_{\theta^*}(\tilde{s})$ as a Gaussian distribution, where the variance is the average received power on K REs.

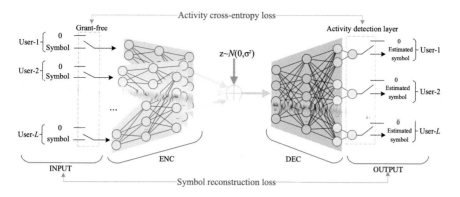

Fig. 7.2 Detailed network structure

However, the remaining obstacle is that the exact value of θ^* is still unknown until now. Hence we replace θ^* with another variational parameter, i.e., θ, as suggested in [31], and solve the following joint optimization problem to achieve the optimal detection accuracy,

$$\min_{\theta,\phi} \mathbb{E}_{p(x)} [-\mathcal{L}(\theta, \phi|x)], \text{ s.t. } \theta \in \Theta, \phi \in \Phi, \tag{7.6}$$

where Θ and Φ are the value spaces of θ and ϕ, respectively. To solve (7.6), we then parameterizes $p_\theta(\cdot)$ and $q_\phi(\cdot)$ with a neural network model of grant-free NOMA to exploit the global function approximation property of DNN and in this case, Θ and Φ are the real vector spaces.

To couple with the system model of grant-free NOMA, we propose a specially designed end-to-end network model, as shown in Fig. 7.2, which contains the characteristics of grant-free NOMA. The proposed network consists of a NOMA encoding network and a decoding network, respectively denoted by $f_{\mathbf{W}_f}(\cdot)$ and $g_{\mathbf{V}_g}(\cdot)$, where $\mathbf{W}_f$ and $\mathbf{V}_g$ are the network parameter sets corresponding to θ and ϕ, respectively. Analogous to $p_\theta(\cdot)$ and $q_\phi(\cdot)$, $f_{\mathbf{W}_f}(\cdot)$ converts the modulated symbols to symbol sequences and $g_{\mathbf{V}_g}(\cdot)$ regenerates the modulated symbols out of the noise-polluted symbol sequences. The details of the proposed network model are presented as follows.

7.3.2 Encoding Network

Different from the conventional VAE which employs a fully-connected neural network at the encoder, the proposed encoding network of DL-aided grant-free NOMA is composed by L isolated sub-networks, since the users cannot obtain external information from other users before transmission.

The encoding sub-network of user-l, denoted as $f_{l,\mathbf{W}_{f_l}}(\cdot)$, maps $\mathbf{x}_l$ to $\mathbf{s}_l$ and is defined as

$$s_l = f_{l,\mathbf{W}_{f_l}}(\mathbf{x}_l) = \tilde{f}_{l,\mathbf{W}_{f_l}}(\mathbf{x}_l) - \tilde{f}_{l,\mathbf{W}_{f_l}}(\mathbf{0}), \tag{7.7}$$

where

$$\begin{aligned}
\tilde{f}_{l,\mathbf{W}_{f_l}}(\mathbf{x}_l) = \sigma_f^{(T)}(\mathbf{W}_l^{(T)}(\sigma_f^{(T-1)} \cdots \sigma_f^{(2)}(\mathbf{W}_l^{(2)} \\
\sigma_f^{(1)}(\mathbf{W}_l^{(1)}\mathbf{x}_l + \boldsymbol{b}_l^{(1)}) + \boldsymbol{b}_l^{(2)}) \cdots + \boldsymbol{b}_l^{(T-1)}) + \boldsymbol{b}_l^{(T)}),
\end{aligned} \tag{7.8}$$

and T is the number of hidden layers, $\mathbf{W}_l^{(t)} \in \mathbb{R}^{d_t \times d_{t+1}}$, $t = 1 \cdots T$, denotes the parameter matrix associated with the connections between layer-t and layer-$(t + 1)$, d_t denotes the number of neurons in layer-t, $\boldsymbol{b}^{(t)}$ denotes the bias associated with layer-$(t + 1)$, and $\sigma_f^{(t)}$ is the activation function of layer-t, which can be either Sigmoid, Tanh, Softmax, or rectified linear unit (ReLU) functions [30]. Besides, $\mathbf{W}_{f_l} = \{\mathbf{W}_l^{(1)}, \boldsymbol{b}_l^{(1)}, \cdots, \mathbf{W}_l^{(T)}, \boldsymbol{b}_l^{(T)}\}$ represents the parameter set of $f_{l,\mathbf{W}_{f_l}}(\cdot)$. We note that, (7.7) performs a centralization to ensure that the transmit signal of inactive users is $\mathbf{0}$. Generally, applying (7.8) results in non-linear symbol spreading where the spread sequences of different symbols are linearly independent. To derive the linear spreading signatures, we can assume $T = 1$ and use identity function as the activation function.

The outputs of the L sub-networks are then added together to derive the composite symbol sequence, as follows

$$f_{\mathbf{W}_f}(\mathbf{x}) = \sum_{l=1}^{L} \sqrt{P_l} \, \mathrm{diag}(\boldsymbol{h}_l) \, f_{l,\mathbf{W}_{f_l}}(\mathbf{x}_l), \tag{7.9}$$

where $\mathbf{W}_f = \{\mathbf{W}_{f_1}, \cdots, \mathbf{W}_{f_l}, \cdots, \mathbf{W}_{f_L}\}$. Ultimately, $q_\phi(\tilde{s}|\mathbf{x})$ is approximated by

$$q_\phi(\tilde{s}|\mathbf{x}) \leftarrow \mathcal{N}(\tilde{s}|f_{\mathbf{W}_f}(\mathbf{x}), \sigma_0^2 \mathbf{I}). \tag{7.10}$$

7.3.3 Decoding Network

In uplink NOMA, the receiver aims to jointly detect L signal streams. Therefore, a fully-connected neural network can be deployed as the probabilistic decoder. In this chapter, we consider the decoding network with multivariate Gaussian outputs [31], which is given as follows,

$$p_\theta(\mathbf{x}|\tilde{s}) \leftarrow \mathcal{N}(\mathbf{x}|g_{\mathbf{V}_g}(\tilde{s}), \sigma_0^2 \mathbf{I}), \tag{7.11}$$

where

$$
\begin{aligned}
g_{\mathbf{V}_g}(\tilde{s}) = \sigma_g^{(R)}(\mathbf{V}^{(R)}(\sigma_g^{(R-1)}\cdots\sigma_g^{(2)}(\mathbf{V}^{(2)} \\
\sigma_g^{(1)}(\mathbf{V}^{(1)}\tilde{s} + c^{(1)}) + c^{(2)})\cdots + c^{(R-1)}) + c^{(R)}),
\end{aligned}
\tag{7.12}
$$

and R is the number of hidden layers, $\mathbf{V}^{(r)} \in \mathbb{R}^{d_r \times d_{r+1}}$, $r = 1\cdots R$, denotes the parameter matrix associated with the connections between layer-r and layer-$(r+1)$, d_r denotes the number of neurons in layer-r, $c^{(r)}$ denotes the bias associated with layer-$(r+1)$, and $\sigma^{(r)}$ is the activation function of layer-r. Besides, $\mathbf{V}_g = \{\mathbf{V}^{(1)}, c^{(1)}, \cdots, \mathbf{V}^{(R)}, c^{(R)}\}$ represents the parameter set of $g_{\mathbf{V}_g}(\cdot)$.

Due to the grant-free access mechanism, the receiver shall also identify the user activity $\mathcal{I}$ according to $\tilde{s}$. Intuitively, the reconstructed symbols, i.e., $g_{\mathbf{V}_g}(\tilde{s})$, hold the information about whether a user is activated. To this end, we just add a single layer with Bernoulli output after $g_{\mathbf{V}_g}(\tilde{s})$ to detect the user activity with ultra-low complexity, as follows

$$
\log p(\mathcal{I}|\tilde{s}) = \mathcal{I}^T \log \hat{\mathcal{I}} + (1 - \mathcal{I}^T)\log(1 - \hat{\mathcal{I}})
\tag{7.13}
$$

where,

$$
\hat{\mathcal{I}} = \sigma^{(\mathcal{I})}(\mathbf{V}^{(\mathcal{I})}g_{\mathbf{V}_g}(\tilde{s}) + c^{(\mathcal{I})}),
\tag{7.14}
$$

where $\sigma^{(\mathcal{I})}$ is the Sigmoid or Tanh activation function, and $\mathbf{V}_{\mathcal{I}} = \{\mathbf{V}^{(\mathcal{I})}, c^{(\mathcal{I})}\}$ is the parameter of this layer. A simple hard decision with threshold $1/2$ is then made on $\hat{\mathcal{I}}$ to get the estimated activity state of the users.

7.4 Multi-loss Based Network Training Algorithm

In the last section, we have proposed the network model for DL-aided grant-free NOMA. To achieve detection accuracy gain over conventional grant-free NOMA, the key is to properly train this network. In this section, we first present the training dataset, and then design a novel loss function, which exploits the prior information about user activity probability. The detailed training algorithm is discussed afterwards.

7.4.1 Dataset Organization with Random User Activation

We adopt the supervised learning method to train the DL-aided grant-free NOMA. Different from the conventional DL tasks, i.e., CV and NLP, where the underlaying distributions of the raw data do not have explicit forms, the raw data samples processed in the proposed network are exactly the randomly generated modulation symbols at the transmitters according to the predefined constellation alphabets and the grant-free access mechanism. Therefore, it is sufficient to use a synthetic dataset to train the proposed network. We denote the dataset as $\mathcal{X} = \{x^{(i)}, \mathcal{I}^{(i)}\}_{i=1}^{N}$, where N is the size of samples, $x^{(i)} = [(x_1^{(i)})^\mathrm{T}, \cdots, (x_l^{(i)})^\mathrm{T}, \cdots, (x_L^{(i)})^\mathrm{T}]^\mathrm{T}$, and $x_l^{(i)}$ is the i.i.d. sample of x_l according to (7.3). Besides, $\mathcal{I}^{(i)}$ is the user activity indicating vector according to $x^{(i)}$. $\mathcal{X}$ is further randomly divided into a training set $\mathcal{X}^{\mathrm{Train}}$ and a testing set $\mathcal{X}^{\mathrm{Test}}$ for evaluation purpose.

7.4.2 Multi-loss Function Design

The loss function compares the differences between estimated output and real output. Through minimizing the loss function, the parameters are fine-tuned to suit the learning task. Hence, the design of loss function greatly affects the performance of deep learning. The total loss of DL-aided grant-free NOMA shall consist of multiple loss functions including modulation symbol reconstruction loss and user activity detection loss.

7.4.2.1 Reconstruction Loss

For the purpose of reconstructing the transmitted symbol, we derive a loss function based on (7.5)

$$
\begin{aligned}
\text{(Loss-A): } &\tilde{\mathcal{L}}^A(\mathbf{W}_f, \mathbf{V}_g | x^{(i)}) \\
&= \mathbb{E}_{\tilde{s} \sim \mathcal{N}(\tilde{s}|f_{\mathbf{W}_f}(x^{(i)}), \sigma_0^2 I)} \left[\log \mathcal{N}(x^{(i)} | g_{\mathbf{V}_g}(\tilde{s}), \sigma_0^2 I) \right] \\
&\quad - \mathrm{KL}(\mathcal{N}(\tilde{s}|f_{\mathbf{W}_f}(x^{(i)}), \sigma_0^2 I) || \mathcal{N}(\tilde{s}|0, P_{\mathrm{ave}} I)) \\
&= \tilde{\mathcal{L}}^{A_1}(\mathbf{W}_f, \mathbf{V}_g | x^{(i)}) + \tilde{\mathcal{L}}^{A_2}(\mathbf{W}_f | x^{(i)}),
\end{aligned}
\tag{7.15}
$$

where P_{ave} is the average received power per dimension per RE.

To efficiently calculate $\tilde{\mathcal{L}}^A(\mathbf{W}_f, \mathbf{V}_g | x^{(i)})$, the first RHS term can be estimated with the repremeterization trick [31], as follows

$$
\begin{aligned}
&\tilde{\mathcal{L}}^{A_1}(\mathbf{W}_f, \mathbf{V}_g | x^{(i)}) \\
&\quad \approx \sum_{s=1}^{S} \log \mathcal{N}(x^{(i)} | g_{\mathbf{V}_g}(f_{\mathbf{W}_f}(x^{(i)}) + \epsilon^{(s)}), \sigma_0^2 I), \\
&\quad \epsilon^{(s)} \sim \mathcal{N}(0, \sigma_0^2 I),
\end{aligned}
\tag{7.16}
$$

and the second RHS term is analytically derived as

$$
\begin{aligned}
\tilde{\mathcal{L}}^{A_2}(\mathbf{V}_g | \mathbf{x}^{(i)}) = & K \left(1 + \log \frac{\sigma_0^2}{P_{\text{ave}}} - \frac{\sigma_0^2}{P_{\text{ave}}} \right) \\
& - \frac{1}{\cdot} \left(f_{\mathbf{W}_f}(\mathbf{x}^{(i)}) \right)^{\mathsf{T}} f_{\mathbf{W}_f}(\mathbf{x}^{(i)}).
\end{aligned}
\tag{7.17}
$$

7.4.2.2 Activity Detection Loss with Confidence Penalty

Although minimizing $\tilde{\mathcal{L}}^A(\mathbf{W}_f, \mathbf{V}_g | \mathbf{x}^{(i)})$ ensures the symbol reconstruction accuracy, a criterion is still required to determine the user activation. Here we use the cross-entropy loss for binary classification, as follows,

$$
\begin{aligned}
& \tilde{\mathcal{L}}^{B_1}(\mathbf{W}_f, \mathbf{V}_g, \mathbf{V}_I | \mathbf{x}^{(i)}, \mathcal{I}^{(i)}) \\
& = \mathbb{E}_{\mathcal{N}(\tilde{s} | f_{\mathbf{W}_f}(\mathbf{x}^{(i)}), \sigma_0^2 I)} \left[\log p(\mathcal{I}^{(i)} | \tilde{s}) \right]
\end{aligned}
\tag{7.18}
$$

where $\log p(\mathcal{I}^{(i)} | \tilde{s})$ is calculated based on (7.13).

For IoT missions, sometimes the activity probability of the users may be close to zero, hence the prior distribution of user activity is usually far away from the uniform distribution. Under such circumstances, there are occasions that the proposed network model over-concentrates on the category with large probability [37], and thus deviates from the true distribution of user activity. To exploit the prior information of user activity, we incorporate a confidence penalty to penalize the cases where the estimated distributions of user activity does not obey the prior distribution, as follows

$$
\begin{aligned}
& \tilde{\mathcal{L}}^{B_2}(\mathbf{W}_f, \mathbf{V}_g, \mathbf{V}_I | \mathbf{x}^{(i)}, \mathcal{B}(\alpha)) \\
& = \mathbb{E}_{\tilde{s} \sim \mathcal{N}(\tilde{s} | f_{\mathbf{W}_f}(\mathbf{x}^{(i)}), \sigma_0^2 I)} \left[\mathrm{KL}(\mathcal{B}(\alpha) || p(\mathcal{I}^{(i)} | \tilde{s})) \right]
\end{aligned}
\tag{7.19}
$$

where $\mathcal{B}(\alpha)$ is the Bernoulli distribution with parameter α. Noticing that, $\mathrm{KL}(\mathcal{B}(\alpha) || p(\mathcal{I}^{(i)} | \tilde{s}))$ can be simplified as

$$
\mathrm{KL}(\mathcal{B}(\alpha) || p(\mathcal{I}^{(i)} | \tilde{s})) = \mathrm{H}(\mathcal{B}(\alpha), p(\mathcal{I}^{(i)} | \tilde{s})) - \mathrm{H}(\mathcal{B}(\alpha)),
\tag{7.20}
$$

where $\mathrm{H}(\mathcal{P}, \mathcal{Q})$ is the cross entropy function, and $\mathrm{H}(\mathcal{P})$ is the differential entropy function. Without loss of generality, we omit $\mathrm{H}(\mathcal{B}(\alpha))$ from the loss function since it is a constant value. Hence, we merge (7.18) and (7.19) and derive the complete loss function of activity detection, as follows

$$(\text{Loss-B}) : \tilde{\mathcal{L}}^B(\mathbf{W}_f, \mathbf{V}_g, \mathbf{V}_I | \boldsymbol{x}^{(i)}, \mathcal{I}^{(i)}, \mathcal{B}(\alpha))$$
$$= (1 - e)\tilde{\mathcal{L}}^{B_1}(\mathbf{W}_f, \mathbf{V}_g, \mathbf{V}_I | \boldsymbol{x}^{(i)}, \mathcal{I}^{(i)})$$
$$+ e\tilde{\mathcal{L}}^{B_2}(\mathbf{W}_f, \mathbf{V}_g, \mathbf{V}_I | \boldsymbol{x}^{(i)}, \mathcal{B}(\alpha))$$
$$= \mathbb{E}_{\tilde{s} \sim \mathcal{N}(\tilde{s} | f_{\mathbf{w}_f}(\boldsymbol{x}^{(i)}), \sigma_0^2 I)} \Big[(1 - e) \log p(\mathcal{I}^{(i)} | \tilde{s})$$
$$+ e\mathrm{H}(\mathcal{B}(\alpha), p(\mathcal{I}^{(i)} | \tilde{s})) \Big] \tag{7.21}$$
$$\overset{(a)}{=} \mathbb{E}_{\tilde{s} \sim \mathcal{N}(\tilde{s} | f_{\mathbf{w}_f}(\boldsymbol{x}^{(i)}), \sigma_0^2 I)} \Big[\log p((1 - e)\mathcal{I}^{(i)} + \alpha e | \tilde{s}) \Big]$$
$$\approx \sum_{s=1}^{S} \log p((1 - e)\mathcal{I}^{(i)} + \alpha e | f_{\mathbf{w}_f}(\boldsymbol{x}^{(i)}) + \boldsymbol{\epsilon}^{(s)}),$$
$$\boldsymbol{\epsilon}^{(s)} \sim \mathcal{N}(\mathbf{0}, \sigma_0^2 I),$$

where $0 \le e \le 1$ is the hyper-parameter controlling the weight of confidence penalty, and step (a) is due to

$$(1 - e) \log p(\mathcal{I}^{(i)} | \tilde{s}) + e\mathrm{H}(\mathcal{B}(\alpha), p(\mathcal{I}^{(i)} | \tilde{s}))$$
$$= \sum_{l=1}^{L} (1-e)(\mathcal{I}^{(i)}(l) \log \hat{\mathcal{I}}^{(i)}(l) + (1 - \mathcal{I}^{(i)}(l)) \log(1 - \hat{\mathcal{I}}^{(i)}(l)))$$
$$+ e(\alpha \log \hat{\mathcal{I}}^{(i)}(l) + (1 - \alpha) \log(1 - \hat{\mathcal{I}}^{(i)}(l)))$$
$$= \sum_{l=1}^{L} ((1 - e)\mathcal{I}^{(i)}(l) + \alpha e) \log \hat{\mathcal{I}}^{(i)}(l) \tag{7.22}$$
$$+ (1 - ((1 - e)\mathcal{I}^{(i)}(l) + \alpha e)) \log(1 - \hat{\mathcal{I}}^{(i)}(l))$$
$$= \log p((1 - e)\mathcal{I}^{(i)} + \alpha e | \tilde{s}).$$

Now we are ready to propose the multi-loss function given by the weighted sum of (7.15) and (7.21), as follows

$$(\text{Total Loss}) : \tilde{\mathcal{L}}(\mathbf{W}_f, \mathbf{V}_g, \mathbf{V}_I | \boldsymbol{x}^{(i)}, \mathcal{I}^{(i)}, \mathcal{B}(\alpha))$$
$$= \gamma_A \tilde{\mathcal{L}}^A(\mathbf{W}_f, \mathbf{V}_g | \boldsymbol{x}^{(i)}) \tag{7.23}$$
$$+ \gamma_B \tilde{\mathcal{L}}^B(\mathbf{W}_f, \mathbf{V}_g, \mathbf{V}_I | \boldsymbol{x}^{(i)}, \mathcal{I}^{(i)}, \mathcal{B}(\alpha)),$$

where γ_A and γ_B are the weight parameters, and shall be derived empirically.

7.4.3   Overall Algorithm

The standard forward and backward propagations are performed to train the proposed grant-free NOMA network, as illustrated in Algorithm 7.1. During each training iteration, $\mathcal{X}^{\text{Train}}$ is divided into mini-batches with each having N' samples. Then We adopt mini-batch training and gradient descent based optimizer, e.g., Adam optimizer, to update the network parameters. While directly optimizing the total loss at the very start of the training tends to mislead the optimizer, we employ the alternative training method to train the proposed network. As shown in the lines 9–12, when the iteration number is small than a threshold E, only symbol reconstruction loss (7.15) is minimized, while otherwise, the total loss (7.23) is minimized.

Since network training costs a large amount of computing resources, the proposed network model is trained offline based on Algorithm 7.1. After convergence, the L trained encoding sub-networks respectively constitute the spreading signatures of L users, and are then applied in non-orthogonal data transmissions, where the conventional receivers, such as MF receiver, is applied to distinguish different data

---

Algorithm 7.1 Forward-backward training algorithm for DL-aided grant-free NOMA

Require: User activation probability "α", channel coefficient "h_l, $\forall l$", noise variance "σ_0^2", average received power per RE "P_0", and dataset "$\mathcal{X}^{\text{Train}}$". Hyper-parameters: sampling size "S", weight of symbol reconstruction loss "γ_A", weight of user activity detection loss "γ_B", weight of confidence penalty "e", and threshold of alternative training "E".

Ensure: Network parameters: Parameters of NOMA encoding network "$\mathbf{W}_f$", parameters of NOMA decoding network "$\mathbf{V}_g$", and parameters of activity detection layer "$\mathbf{V}_I$".

1: $\mathbf{W}_f, \mathbf{V}_g, \mathbf{V}_I \leftarrow$ Randomly initialize parameters
2: **repeat**
3:     *(Forward propagation)*
4:     $\{x^{(i)}, \mathcal{I}^{(i)}\}_{i=1}^{N'} \leftarrow$ Randomly draw a mini-batch with N' samples out of $\mathcal{X}^{\text{Train}}$
5:     $\tilde{\mathcal{L}}_{\text{Sample}}^{(i)} \leftarrow 0, 1 \leq i \leq N'$
6:     $\tilde{\mathcal{L}}_{\text{Mini-batch}} \leftarrow 0$
7:     $\nabla_{\mathbf{W}_f}, \nabla_{\mathbf{V}_g}, \nabla_{\mathbf{V}_I} \leftarrow 0$
8:     **for** $i \leq N'$ **do**
9:         **if** Iteration number is smaller than E **then**
10:             $\tilde{\mathcal{L}}_{\text{Sample}}^{(i)} \leftarrow \tilde{\mathcal{L}}^A(\mathbf{W}_f, \mathbf{V}_g | x^{(i)})$ in (7.15)
11:         **else**
12:             $\tilde{\mathcal{L}}_{\text{Sample}}^{(i)} \leftarrow \tilde{\mathcal{L}}(\mathbf{W}_f, \mathbf{V}_g, \mathbf{V}_I | x^{(i)}, \mathcal{I}^{(i)}, \mathcal{B}(\alpha))$ in (7.23)
13:         **end if**
14:         $\tilde{\mathcal{L}}_{\text{Mini-batch}} \leftarrow \tilde{\mathcal{L}} + \frac{1}{N'}\tilde{\mathcal{L}}_{\text{Sample}}^{(i)}$
15:         $i \leftarrow i + 1$
16:     **end for**
17:     *(Backward propagation)*
18:     $\nabla_{\mathbf{W}_f}, \nabla_{\mathbf{V}_g}, \nabla_{\mathbf{V}_I} \leftarrow \nabla_{\mathbf{W}_f, \mathbf{V}_g, \mathbf{V}_I} \tilde{\mathcal{L}}_{\text{Mini-batch}}$ (Calculate gradient for parameters)
19:     $\mathbf{W}_f, \mathbf{V}_g, \mathbf{V}_I \leftarrow$ Update parameters based on $\nabla_{\mathbf{W}_f}, \nabla_{\mathbf{V}_g}, \nabla_{\mathbf{V}_I}$ using, e.g., SGD or Adam
20: **until** Convergence of $\mathbf{W}_f, \mathbf{V}_g$, and $\mathbf{V}_I$.

---

streams. We note that, during data transmission, our proposed scheme does not lead to any increase in computational complexity or latency compared with conventional grant-free NOMA.

7.5 Simulation Results

In this section, experiments and Monte-Carlo simulations are conducted to validate the performance gain of the proposed DL-aided grant-free NOMA. The proposed network is constructed and trained based on the well-known Tensorflow framework. A GPU server with two Nvidia® GTX-1080Ti GPUs and a Intel® Xeon E5-2620 v4 CPU is employed to accelerate the training procedure, while the proposed network can also be trained on an ordinary consumer computer. The network parameters are initialized with Xavier initialization method.

7.5.1 Network Training Results and Design Examples

The settings of the proposed network are as follows. For each user, the NOMA encoding network consists of one input layer and four hidden layers with M, 16, 32, 16, and 8 neurons, respectively. The NOMA decoding network consists of four hidden layers, one reconstructed symbol output layer, and one activity estimation output layer, with 64, 128, 128, 32, $M \times L$, and L neurons, respectively. Sigmoid function is utilized as the activation function of all layers, except the output layer of NOMA encoding network where Tanh function is used to generate symmetrical outputs. We use $\mathfrak{X}^{\text{Train}}$ to train the proposed network model.

Figure 7.3 shows the total loss during training versus iteration number with or without alternative training, where the learning rate is set as 10^{-4}, the batch size is set as 256, and other hyper-parameters are fine-tuned in different simulation settings with grid search. We assume $L = 6$ and $K = 4$, and the model is trained under 1.5 dB per user. As an initial study, AWGN channel with constant channel coefficients is assumed [35]. The detailed hyper-parameter settings are as follows: $S = 2$, $\gamma_A = 1$, $\gamma_B = 0.5$, $e = 0.1$ and $E = 1000$. After convergence, smaller total loss is achieved by alternative training, which results in a slightly lower deviation on symbol reconstruction as well as a slightly higher accuracy on user activity estimation. Besides, the network is randomly initialized and trained until convergence for several times, which validates the stability of the proposed training algorithm.

The proposed network is trained offline, and the training results of the NOMA encoding network are the learned mapping relationships between x_l and s_l, which are applied in grant-free NOMA. As discussed earlier, the proposed network usually learns non-linear symbol spreading signatures, i.e., different modulation symbols correspond to different spreading signatures, which can be illustrated by multi-

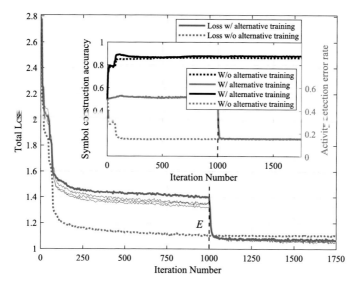

Fig. 7.3 Total loss, symbol error rate, and activity detection accuracy vs iteration number, with or without alternative training

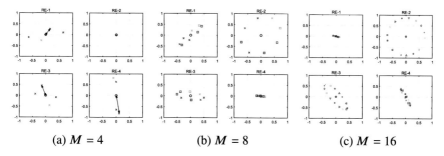

(a) *M* = 4   (b) *M* = 8   (c) *M* = 16

Fig. 7.4 Examples of the learned multi-dimensional constellations of one user in 4 REs with different M. In each example, the 4 points with the same color and shape in 4 REs constitute a four-dimensional constellation point

dimensional constellations where each source message corresponds to a multi-dimensional constellation point [38].

Figure 7.4 illustrates some examples of the learned multi-dimensional constellations with various message sizes and activation probabilities. For each constellation, the points in the same color and shape in 4 REs constitute a multi-dimensional constellation point. Some interesting phenomena are observed from the results. First of all, *constellation projection* operation [39], where the multi-dimensional constellation points of different messages may overlap on some REs, is naturally learned in the proposed scheme, as shown in RE-2 of Fig. 7.4a, RE-4 of Fig. 7.4b, RE-1 and RE-4 of Fig. 7.4c. We note that the constellation projection has been shown to reduce the receiving complexity [39]. Besides, the learned constellation tends to transmit signals with various power levels in 4 REs, which inherently introduce power gain differences, and can facilitate the interference cancellation. An example is highlighted in Fig. 7.4a, where the arrows reflect transmission powers of the black multi-dimensional constellation point in different REs. A sparse RE mapping manner [14] is also learned, as shown in Fig. 7.4a, which is proven to reduce the receiving complexity. These observations show that DL-aided grant-free NOMA automatically tap into some physical insights on the design of spreading signature, which conventionally requires expert knowledge in wireless communications. We note that the learned constellations usually have irreguler shapes, and differ from the conventional constellations, e.g. M-QAM, and M-PAM, which may confuse the potential eavesdroppers and help enhance the transmission security in Tactile IoT.

While multi-dimensional constellation normally requires iterative-based detection methods [10] which dissatisfy the low latency requirement of Tactile IoT, we apply a linear constraint to derive linear spreading signatures for grant-free NOMA, where all modulation symbols of a user have the same spreading signature. This helps to implement simple detectors, e.g. match-filter detection. With the setting of $L = 6$, $K = 4$, $M = 4$ and QPSK modulation, a randomly generated example of the learned spreading signature set is shown in (7.24), where the l-th column is the normalized spreading signature of user-l. Arbitrary number of spreading signature sets can be generated automatically with different settings, which suits the diversified and highly automatic communications in the future Tactile IoT.

$$S_{4 \times 6} = \begin{bmatrix} -0.08 + 0.23i & -0.20 - 0.20i & -0.01 - 0.23i & 1.37 - 1.33i & 0.97 - 0.27i & 0.68 + 0.01i \\ -0.50 - 0.47i & 0.26 + 0.55i & -1.06 - 0.40i & 0.54 + 0.34i & 0.48 - 0.66i & -0.12 - 0.25 \\ -0.32 - 0.22i & -0.23 + 0.07i & 1.66 + 0.18i & 0.25 - 0.14i & 0.27 - 0.62i & -1.36 + 0.19i \\ -0.52 - 0.19i & -0.52 - 0.47i & 0.13 + 0.12i & -0.08 + 0.63i & -0.45 - 0.54i & 1.15 - 0.44i \end{bmatrix}$$

$$(7.24)$$

7.5.2 Detection Accuracy Analysis

In this subsection, we evaluate the symbol error rate (SER) and activity detection accuracy of the proposed scheme using $\mathcal{X}^{\text{Test}}$. As shown in Fig. 7.5, we first compare the achievable SER of the linear spreading signatures generated by the proposed scheme with state-of-art grant-free NOMA schemes, i.e., MUSA and WBE [40], under a simple MF receiver. It is observed that, about 5dB gain in SER can be achieved compared to conventional grant-free NOMA while not introducing any increase in receiving complexity or latency. We then employ the DNN-based receiver (DNNrx), which constitutes a near maximum-likelihood (ML) receiver and is trained along with the linear spreading signatures. In this case, 10^{-3} SER can be achieved with various activation probabilities, which is promising to satisfy the rigorous requirement on the reliability in Tactile IoT. Although DNNrx leads to the increase of computation complexity compared with MF receiver, DNNrx, as a non-iterative receiver, causes

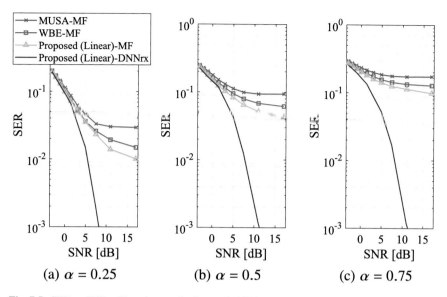

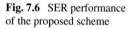

(a) $\alpha = 0.25$ (b) $\alpha = 0.5$ (c) $\alpha = 0.75$

Fig. 7.5 SER vs SNR with various activation probabilities

Fig. 7.6 SER performance of the proposed scheme

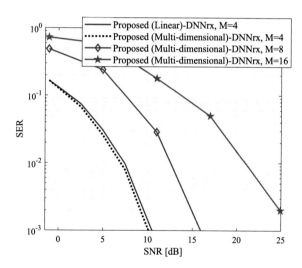

much less detection latency compared with other iterative-based near ML receiver, e.g., message passing algorithm-based receivers. This makes DNNrx still feasible for Tactile IoT.

The SER performances of the propose scheme with various M is shown in Fig. 7.6. To evaluate the ultimate reliability performance of DL-aided grant-free NOMA, here we adopt a larger network than that in Sect. 7.5.1, where the largest layers at NOMA encoding and decoding consist of 64 and 512 neurons, respectively. It is observed

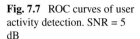 **Fig. 7.7** ROC curves of user activity detection. SNR = 5 dB

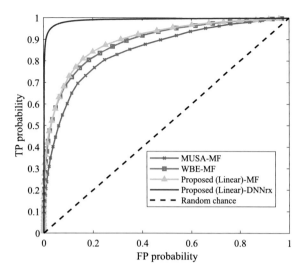

that ultra-high reliability is achieved with various M, which validates the universality of the proposed scheme. Besides, we observe that multi-dimensional signatures achieve slightly better SER performance than linear signatures with DNNrx. This is because the multi-dimensional signatures are more flexible than its linear counterparts. However, maximum-likelihood based detection algorithms are required to distinguish multiple data streams with multi-dimensional signatures, which leads to larger detection delay. On the contrary, the overlapped signals with linear spreading signatures can be quickly separated even with a very simple receiver, e.g. MF receiver, which is more suitable for Tactile IoT.

Figure 7.7 compares the user activity detection performance of the proposed spreading signatures (7.24) with conventional grant-free NOMA schemes. To detect the activity of the users, MF receiver is first applied to suppress the inter-user interference and then energy detection is employed for the final decision. The receiver operating characteristic (ROC) curve, which reflects the relationship between $\mathcal{E}_{TP}$ and $\mathcal{E}_{FP}$, is presented for performance evaluation, where each point on the curve represents an arbitrary decision threshold in energy detection. The results show that the proposed scheme achieves higher accuracy than conventional schemes, while utilizing the proposed NOMA encoding network can further enhance the accuracy. The reliability gain of the proposed scheme with respect to user activity detection and symbol reconstruction can reduce re-transmissions number and lead to further latency reduction.

7.6 Conclusions

In this chapter, a DL-aided grant-free NOMA scheme was proposed to tackle the challenge of establishing ultra-responsive and ultra-reliable connectivities for massive devices in Tactile IoT. A neural network model of grant-free NOMA as well as the corresponding multi-loss function was constructed to exploit the grant-free behavior in the design of symbol spreading signatures. The proposed scheme avoids the complicated human-crafted work and enables the automatic design on spreading signatures, which ideally suits the highly automatic applications in the future Tactile IoT. Simulation results showed that the proposed scheme achieves higher reliability than the conventional grant-free NOMA schemes.

There are several important extensions for further evolution of Tactile IoT based on DL. First off, the historical data can be analyzed by deep recurrent models to exploit the activation patterns of devices. Besides, the diversified requirements of IoT, such as low power consumption and high reliability, shall be considered in the design of radio access schemes via multi-task learning approach. More importantly, the vast amount of high-dimensional raw data in the network layer of Tactile IoT shall be exploited by deep representation learning to integrally enhance the network functions as well as radio interface, which may finally lead to a systematic paradigm shift in the communication part of Tactile IoT.

References

1. A. Konstantinos et al., Towards haptic communications over the 5G Tactile Internet. IEEE Commun. Surv. Tutor. **20**(4), 3034–3059 (2018)
2. M. Martin et al., The tactile internet: vision, recent progress, and open challenges. IEEE Commun. Mag. **54**(5), 138–145 (2016)
3. Q. Zhang, et al., Mission critical IoT communication in 5G, in *Future Access Enablers for Ubiquitous and Intelligent Infrastructures* (2015), pp. 35–41
4. A. Aijaz et al., Realizing the tactile internet: haptic communications over next generation 5G cellular networks. IEEE Wirel. Commun. **24**(2), 82–89 (2017)
5. G.P. Fettweis, The tactile internet: applications and challenges. IEEE Veh. Technol. Mag. **9**(1), 64–70 (2014)
6. M. Simsek et al., 5G-enabled tactile internet. IEEE J. Sel. Areas Commun. **34**(3), 460–473 (2016)
7. K. Niu et al., Polar codes: primary concepts and practical decoding algorithms. IEEE Commun. Mag. **52**(7), 192–203 (2014)
8. M. Iwabuchi, et al., Evaluation of coverage and mobility for URLLC via outdoor experimental trials, in *Proceedings of the IEEE 87th Vehicular Technology Conference (VTC-Spring)* (IEEE, Porto, Portugal, 2018), pp. 1–5
9. X. Li et al., A dynamic decision-making approach for intrusion response in industrial control systems. IEEE Trans. Ind. Inform. **15**(5), 2544–2554 (2019)
10. N. Ye et al., Uplink non-orthogonal multiple access technologies toward 5G: a survey, in *Wireless Communications and Mobile Computing* (2018)

11. G. Ma et al., Coded tandem spreading multiple access for massive machine-type communications. IEEE Wirel. Commun. **25**(2), 75–81 (2018)
12. Y. Du et al., Efficient multi-user detection for uplink grant-free NOMA: prior-information aided adaptive compressive sensing perspective. IEEE J. Sel. Areas Commun. **35**(12), 2812–2828 (2017)
13. 3GPP, R1-1713952, UL data transmission without UL grant (2017)
14. H. Yu et al., Optimal design of resource element mapping for sparse spreading non-orthogonal multiple access. IEEE Wirel. Commun. Lett. **7**(5), 744–747 (2018)
15. K. Xiao et al., On capacity-based codebook design and advanced decoding for sparse code multiple access systems. IEEE Trans. Wirel. Commun. **17**(6), 3834–3849 (2018)
16. N. Ye et al., On constellation rotation of NOMA with SIC receiver. IEEE Commun. Lett. **22**(3), 514–517 (2018)
17. X. Shao, et al., Dynamic IoT device clustering and energy management with hybrid NOMA systems. IEEE Trans. Ind. Inform. (2018)
18. S. Hu, et al., Non-orthogonal interleave-grid multiple access scheme for industrial Internet-of-Things in 5G network. IEEE Trans. Ind. Inform. (2018)
19. S. Shamai et al., Information-theoretic considerations for symmetric, cellular, multiple-access fading channels. II. IEEE Trans. Inf. Theory **43**(6), 1895–1911 (1997)
20. P. Minero et al., Random access: an information-theoretic perspective. IEEE Trans. Inf. Theory **58**(2), 909–930 (2012)
21. Z. Yuan, et al., Blind multiple user detection for grant-free MUSA without reference signal, in *Proceedings of the IEEE 86th Vehicular Technology Conference (VTC-Fall)* (Toronto, Canada, 2017), pp. 1–5
22. J. Zhang et al., PoC of SCMA-based uplink grant-free transmission in UCNC for 5G. IEEE J. Sel. Areas Commun. **35**(6), 1353–1362 (2017)
23. X. Dai et al., Pattern division multiple access: a new multiple access technology for 5G. IEEE Wirel. Commun. **25**(2), 54–60 (2018)
24. A. Azari, et al., Grant-free radio access for short-packet communications over 5G networks, in *Proceedings of the IEEE Global Communications Conference (GLOBECOM)* (Singapore, 2017), pp. 1–7
25. N. Ye et al., Rate-adaptive multiple access for uplink grant-free transmission. Wirel. Commun. Mob. Comput. **2018**, Article ID 8978207, 21 (2018)
26. B. Wang et al., Dynamic compressive sensing-based multi-user detection for uplink grant-free NOMA. IEEE Commun. Lett. **20**(11), 2320–2323 (2016)
27. A.C. Cirik et al., Multi-user detection using ADMM-based compressive sensing for uplink grant-free NOMA. IEEE Wirel. Commun. Lett. **7**(1), 46–49 (2018)
28. C. Wei et al., Approximate message passing-based joint user activity and data detection for NOMA. IEEE Commun. Lett. **21**(3), 640–643 (2017)
29. Z. Zhang et al., Grant-free rateless multiple access: a novel massive access scheme for Internet of Things. IEEE Commun. Lett. **20**(10), 2019–2022 (2016)
30. Y. LeCun et al., Deep learning. Nature **521**(7553), 436–444 (2015)
31. D.P. Kingma, M. Welling, Auto-encoding variational Bayes, in *Proceedings of the ICLR* (Banff, Canada, 2014), pp. 1–14
32. T. Wang et al., Deep learning for wireless physical layer: opportunities and challenges. China Commun. **14**(11), 92–111 (2017)
33. X. You, et al., AI for 5G: research directions and paradigms. Sci. China Inf. Sci. Rev.
34. T. O'Shea et al., An introduction to deep learning for the physical layer. IEEE Trans. Cogn. Commun. Netw. **3**(4), 563–575 (2017)
35. M. Kim et al., Deep learning-aided SCMA. IEEE Commun. Lett. **22**(4), 720–723 (2018)
36. G. Gui et al., Deep learning for an effective non-orthogonal multiple access scheme. IEEE Trans. Veh. Technol. (2018)
37. G. Pereyra, et al., Regularizing neural networks by penalizing confident output distributions, in *Proceedings of the ICLR* (Toulon, France, 2017), pp. 1–12

38. M. Beko, R. Dinis, Designing good multi-dimensional constellations. IEEE Wirel. Commun. Lett. **1**(3), 221–224 (2012)
39. M. Taherzadeh, et al., SCMA codebook design, in *Proceedings of the IEEE 80th Vehicular Technology Conference (VTC-Fall)* (Vancouver, BC, 2014), pp. 1–5
40. 3GPP, R1-1809974, Updated offline summary of transmitter side signal processing schemes for NOMA (2018)

Chapter 8
Summary and Outlook

Abstract This chapter summarizes the book and discusses the future directions of enhancing multiple access technology for ubiquitous networks.

8.1 Summary

In this book, we discuss various aspects to have a thorough view of the multiple access technology. Specifically, in Chaps. 2 and 3, we discuss multiple access issues towards both terrestrial and non-terrestrial scenarios. In Chaps. 4 and 5, we investigate the methodology of signal design for multiple access. In Chaps. 6 and 7, we study how multiple access technology can be enhanced by AI and DL. The specific contributions of this book are as follows.

In Chap. 2, we discuss the practical deployment of multiple access in 5G and beyond. As the promising multiple access technology, NOMA is now undergoing the standardization process in 5G, owing to the superior performance in spectral efficiency, connectivity, and flexibility. We present a comprehensive review on recent progress of multiple access technology in 5G and beyond, especially NOMA technologies. Specifically, we introduce the typical transmission and reception technologies of multiple access, followed by the grant-free multiple access and implementation issues of multiple access towards 5G.

In Chap. 3, we propose a general look on multiple access technology used in non-terrestrial wireless communication systems with a special focus on IoT scenarios. We first present the requirements of non-terrestrial IoT and point out where the existing terrestrial IoT technologies cannot work. Two non-terrestrial IoT scenarios are then introduced, namely satellite IoT and UAV IoT. The key technologies for satellite and UAV IoTs are reviewed separately. In particular, both physical and non-physical layer technologies for multiple access are surveyed for satellite IoT. Finally, we draw a conclusion and give some potential challenges of non-terrestrial IoT.

In Chap. 4, we investigate the constellation rotation technique for enhancing the performance of uplink multiple access network with SIC receiver. We characterize the received signal with GMM, where optimal rotation angle is obtained by maximizing the entropy of GMM. Then we efficiently solve the entropy maximization problem by deriving an optimal closed-form approximate problem via variational approximation. Performance analysis and simulation results show that the proposed scheme achieves larger capacity and lower BER compared with conventional multiple access.

In Chap. 5, we investigate the rate-splitting technique for uplink grant-free multiple access. A rate-adaptive multiple access scheme is proposed to tackle the collision problems caused by the grant-free transmission. At the receiver, the intra- and inter-user SIC receiving algorithm is employed to detect multiple data streams. In the proposed scheme, the users can achieve rate adaption without the prior knowledge of the channel conditions, since that the layers with high protection property can be successfully recovered when the interference is severe, while other layers can take advantage of the channel when the interference is less significant. Finally, theoretical analysis and simulation results validate that the proposed scheme can simultaneously achieve higher average throughput and lower outage performance than its conventional counterpart.

In Chap. 6, we study how AI and DL can help in approaching the performance limit of multiple access system. Specifically, we resort to deep multi-task learning for end-to-end optimization of NOMA, by regarding the overlapped transmissions as multiple distinctive but correlated learning tasks. First of all, we establish a unified multitask DNN framework for NOMA, namely DeepNOMA, and a multi-task balancing technique is then proposed to guarantee fairness among tasks as well as to avoid local optima. To further exploit the benefits of communication-domain expertise, we introduce constellation shape prior and inter-task interference cancellation structure. Detailed experiments and link-level simulations show that higher transmission accuracy and lower computational complexity can be simultaneously achieved by DeepNOMA under various channel models, compared with state-of-the-art.accuracy and lower computational complexity simultaneously under various channel models.

In Chap. 7, we focus on enhancing grant-free multiple access with deep learning in a specific scenario, i.e., tactile Internet of Things (IoT). We formulate a variational optimization problem to improve the reliability of grant-free NOMA. Due to the intractability of this problem, we resort to deep learning by parameterizing the intractable variational function with a specially designed deep neural network, which incorporates random user activation and symbol spreading. The network is trained according to a novel multi-loss function where a confidence penalty based on the user activation probability is considered. The spreading signatures are automatically generated while training, which matches the highly automatic applications in Tactile IoT. The significant reliability gain of our scheme is validated by simulations.

8.2 Future Directions

In this section, we discuss the future directions of enhancing multiple access technology for ubiquitous networks.

Physical Layer Enhancement

The existing NOMA schemes either focus on bit-level or symbol-level operations, which cannot achieve the global optimal designs for non-terrestrial applications. A straightforward idea is to conduct joint design of bit-level and symbol-level operations, e.g., joint design of channel encoding and symbol spreading, where the coding structure is optimized according to multiuser transmission. However, these designs are sensitive to certain channel conditions, and require further enhancement for practical implementation. In addition, signal detection technologies at the receiver are also required to be enhanced. To exploit the coding structure, several joint detection methods have been proposed. However, the existing methods usually require many iterations between symbol detection and channel encoding, which leads to large detection latency. Furthermore, the high complexity and latency of blind detection are still obstacles to the deployment of grant-free transmission. Therefore, simple and uniform design is required to reduce the computational complexity, as well as to maintain high reliability.

Cross Layer Design

Except for the physical layer enhancement, the cross layer design may also play an important role in the future development of multiple access technology. For example, grant-free multiple access, which integrates the access layer protocol, i.e., grant-free protocol, into multiuser transmission, has been a promising technology for mMTC, as mentioned earlier in this chapter. Besides, it is also promising to combine multiple access with other access layer techniques, such as repetition technique or rateless coding, and optimize the repetition number or code structure.

Different from the physical layer technology which is always inspired by the Shannon information theory, it is non-trivial to derive the capacity limits of the channel when access layer is involved, e.g., the exact achievable channel capacity of grant-free multiple access is still an open problem. One possible solution is to formulate the Shannon information-theoretic channel model for cross layer design. For example, grant-free multiple access can be formulated as the random access channel (RAC) which uses the auxiliary receivers to represent different states of user activation in grant-free access. We note that this channel model is similar to the interference channel model, where rate splitting can achieve a good capacity region. With this insight, one may naturally consider the deployment of rate splitting into grant-free multiple access. To this end, elaborate design and optimization are required to enhance the performance of grant-free multiple access with rate splitting, for example, the coding rate and the power allocation coefficient for each splitting layer.

Scenario-Specific Enhancement

Multiple access technology should be studied towards specific application scenarios. As for the satellite massive access scenario, it is observed that the research is mainly conducted under the LoRa system, while the OFDM-based waveform is less studied. In order to achieve the integration of the non-terrestrial and the terrestrial multiple access in future 6G, we need to unify the two systems. In terms of UAV massive access scenario, there is relatively little research on physical layer technology for large dynamic channel and low power consumption. In order to cope with the high dynamic channel of non-terrestrial communication, asynchronous multiple access is also a possible research direction considering the problem of time-frequency asynchrony of multi-user signals caused by large dynamic channels.

Hardware-Constrained Enhancement

Considering that the hardware resources of non-terrestrial platform are severely limited, we can design the NOMA transmitter with low peak average power ratio and high energy efficiency to meet the antenna and power constraints. And the receiver should be designed based on the general platform to realize the multi-user equalization and receiving technology with low complexity and updating. In addition, the neural network can be introduced to optimize the problem that cannot be expressed in closed form. Moreover, online training can be utilized to adapt to dynamic and complicated equivalent channel.

Besides, to solve the problem of frequent switching of user connection relation which arises from the change of network topology caused by the rapid movement of non-terrestrial platform, we can consider simplifying the user access, registration signaling process and corresponding signal design of physical layer at the signaling process level. In addition, in the physical layer, multi-user multi-satellite/multi-station non-orthogonal communication system is also a feasible research direction.

Joint Design with Other Technologies

Although existing multiple access technologies e.g. NOMA, have met the bottleneck on their own, we can consider to incorporate them into other cutting edge technologies to see if there are additional advantages. For example, full duplex (FD) technology, which is expected to increase the spectrum efficiency by two times, can be jointly designed with NOMA. The conventional FD, which considers a point-to-point communication channel, is modeled by two-way channel (TWC) model. We consider a case where multiple users exist in a cell, and both gNB and the users have FD ability. Consequently, gNB and the users are simultaneously transmitting and receiving, so that the entire system can be regarded as a MAC in uplink, as well as a BC in downlink. In this case, the received signal may be exploited to derive the hidden information about the channel condition and to enhance the transmission reliability.

Furthermore, multiple access technology towards ubiquitous networks can be jointly designed with cooperative transmission, such as cell-free technique, and multiple-antenna technology. Also, there have been some initial studies about employing online DL and few-shot DL to increase the flexibility and efficiency of multipl access technology. Although jointly designing NOMA with other technologies seems straightforward, it is still an open question whether the joint design can produce "a whole greater than the sum of its parts".